BRITISH COLUMBIA
BURNING

The worst wildfire season in B.C. history

Author: Bethany Lindsay

Photo editor: Kelly Sinoski

Foreword: Gary Filmon

Copyright 2018 Bethany Lindsay

All rights reserved. No part of this book covered by the copyrights hereon may be reproduced or used in any form or by any means – graphic, electronic, or mechanical – without the prior written permission of the publisher. Any request for photocopying, recording, taping, or information storage and retrieval systems of any part of this book shall be directed in writing to the Canadian Reprography Collective, 379 Adelaide Street, West, Suite M1, Toronto, Ontario, M5V 1S5.

MacIntyre Purcell Publishing Inc.
194 Hospital Rd.
Lunenburg, Nova Scotia
B0J 2C0
(902) 640-3350

www.macintyrepurcell.com
info@macintyrepurcell.com

Printed and bound in Canada by Friesens.

Design and layout: Denis Cunningham
Cover design: Denis Cunningham
Cover photo: British Columbia experienced its worst wildfire season in 2017, with much of the province going up in flames. Ground and aircrews were deployed across the province, including to this spot near Savona, as the wildfires spun out of control. Kirsten Tasker photo

Author photo (back cover): Kristine Cofsky

ISBN: 978-1-77276-090-3

Library and Archives Canada Cataloguing in Publication

Lindsay, Bethany, author British Columbia Burning / Bethany Lindsay.

ISBN 978-1-77276-090-3 (softcover)

  1. Wildfires--British Columbia. 2. Wildfires--British Columbia--Pictorial works. I. Title.

SD421.34.C3L56 2018   634.9'61809711   C2018-900606-4

MacIntyre Purcell Publishing Inc. would like to acknowledge the financial support of the Government of Canada and the Nova Scotia Department of Tourism, Culture and Heritage.

TABLE OF CONTENTS

AN UNPRECEDENTED YEAR
Few could have predicted that 2017 would be a record-breaking wildfire year. The winter was cold, wet and snowy, and the spring melt-off led to fatal floods. But then, the weather shifted... 9

A DAY TO REMEMBER
By early July, central B.C. was dry as a bone and primed to go up in smoke, but firefighters were having a slow start to the season. All of that changed on July 7.. 19

THE EMERGENCY ESCALATES
A new crisis emerged every few days in July. Things became so dire that the entire city of Williams Lake had to be evacuated. And on the edge of the wildfire zone in Clinton, worried residents could only watch and wonder if they were next 29

AN ACT OF DEFIANCE
Though thousands of people were forced from their homes, every community had a few holdouts who defied evacuation orders. Some First Nations had their own reasons for sticking around — and their own plans for protecting their communities. 37

WHEN THINGS GO WRONG
With a situation this out of control, firefighters were bound to make a few miscalculations. And in the communities of Clinton and Pressy Lake, the results of those miscalculations left residents demanding answers from the province. 47

HOW TO FIGHT A FIRE
The technologies for just about every human pursuit have moved forward by leaps and bounds in recent decades, but the essentials of wildfire fighting haven't changed. What really counts in a crisis are the men and women on the ground...55

THE WORST OF TIMES
With weeks left to go in the wildfire season, it was already official: This was the worst year on record in B.C. By mid-August, crews were struggling to control the biggest fire ever, while new fires kept popping up and threatening communities. 63

HOW DID IT COME TO THIS?
B.C. has gone through bad wildfires before, but critics say the lessons from those catastrophes haven't resulted in significant change in how we manage our landscapes and protect our homes. 69

NOT OVER YET
It was early September, and there was still no end in sight. A new corner of the province was suddenly under threat, while the wildfire service was losing crew members as they returned to school. Meanwhile, B.C.'s wildlife was also feeling the heat. 77

SILVER LININGS
Even in a devastating year like this, there were stories of little miracles, human kindness and bravery that could make the darkest days seem just a little bit brighter. 85

TALLYING UP THE DAMAGE
Hundreds of houses destroyed. Precious memories lost. Thousands of hectares of valuable forest scorched. But material goods aren't the only things damaged in a wildfire, and the people affected will be dealing with the emotional aftermath for a while. 91

FIRES OF THE FUTURE
It's not a matter of if but when B.C. will face another season like this one. As the climate changes, the conditions will become more and more ripe for wildfire. But there are things we can do to prepare. 97

Foreword

WILDFIRE! A terrifying and costly event with significant human consequences. Unfortunately, it's Mother Nature's final solution to the neglect and/or mismanagement of our forest resources.

In British Columbia Burning, Bethany Lindsay provides us with a gripping story that illuminates not just the blunt statistics of the havoc wreaked by the worst wildfire season in British Columbia's modern history, but also the human consequences of the disaster on those who lived through it. The anxieties, the frustrations, and the tragedies involved with the loss of homes, memorabilia and cherished possessions is brought home clearly in the words of those who survived the conflagration of 2017. Those stories of sadness and loss evoked memories of the hundreds of presentations I heard during my many public hearings in the B.C. Interior following the dramatic fires of 2003.

Lindsay's story, painted with words and photography, dramatically remind us of the potential consequences of building our homes and communities in close proximity to our beautiful forests and open spaces — an ironic blending of beauty and risk. In the end, and most importantly, in 2017 no lives were lost and although many of the family heirlooms, photos and letters were irreplaceable, their loss was still less devastating than the loss of human lives would have been.

Lindsay also leaves us with an ominous warning. British Columbia, like California, Australia, New Zealand and large areas of South America, Asia and Europe will face those risks and consequences to an ever-increasing extent as a result of the effects of global climate change on our environment.

As always, life is about choices and this worrisome future will require us to make difficult and gut wrenching choices, both personally and as well a society, such as:

Where do we choose to live in relation to our beautiful and alluring forests and open spaces?

What materials do we choose to build our houses with?

Would it be better to manage our forests with regular controlled burning or leave the decisions to the indiscriminate hand of Mother Nature and the uncontrollable forces of wind and heat that she commands?

Bethany Lindsay's book leaves us with much to consider.

— The Honourable Gary Filmon, P.C., O.C., O.M., LLD
Author of *Firestorm 2003: Provincial Review*

INTRODUCTION

(Right) Randy Thorne, along with his wife Angie (front left), daughter Kelsey Thorne and granddaughter Nevaeh Porter, 8, lost everything as wildfire ravaged their home on the Ashcroft Indian Band Reserve.

(Above) A teapot and mug were rescued from the rubble of a home on the Ashcroft Indian Band Reserve.

(Above) Shawn Cahill, who owns a cabin on Loon Lake, loaded up his possessions and was ready to flee the moment the Elephant Hill wildfire got too hot to handle in the lakeside community.

(Opposite) Thick plumes of smoke erupted over Ashcroft in early July, marking the start of the Elephant Hill fire and a summer few wouldn't easily forget.

When Angie Thorne saw the thick, black smoke spilling into the sky above the Ashcroft Indian Band reserve on July 7th, she knew she wasn't going to make it the rest of the way into Kamloops to get her shopping done.

The wind was blowing fast and from where she was standing outside of the band office, it looked like the wildfire that had been smouldering in the hills overnight had come to life and was headed straight for her house.

"Oh shit," she remembers saying, "this fire is starting to take off."

She called her husband Randy and told him to get his butt home from work, then drove up the hill to her place to get the essentials packed up in case they needed to flee. As the smoke pushed closer and closer, Thorne hooked up her fifth wheel trailer, ran into the house and began grabbing clothes.

But all too soon, it was clear that she was out of time. The choice was to leave now, or risk her life. "I looked around my house and nothing else seemed to matter at that time — just us getting out," she says. Thorne hopped in her car and made a quick detour to make sure her dad and his wife were on their way to safety. She then drove back toward the band office, watching as the RCMP arrived to block the road. Now there was no denying the gravity of the situation.

Thorne had her eight-year-old granddaughter Nevaeh beside her, along with her brother and her dad, but she still hadn't seen her husband. Thorne tried calling and calling to see if he was safe, but Randy wasn't answering and her cell service kept cutting out. She pleaded with the police officers to let her back through the roadblock so she could find him.

Reluctantly, one of the Mounties let her through, but first he warned her: "There's a rule here. If you see flames, get out." Thorne and a friend hopped back in the car. Ignoring the officer's advice, they drove straight past the flames.

Back at her house, the smoke was thicker than ever. She found Randy standing on the roof, with a hose — a valiant last-ditch effort to soak the shingles and maybe save their house — and yelled at him to come down. "I was screaming at him at the top of my lungs," she says. He resisted at first, then finally relented.

Once he was down, they grabbed their vehicles and the rest of their family and drove four kilometres to the town of Ashcroft, where they found a safe place to park. Thorne and a small group of loved ones found a spot on high ground where they could watch the blaze approach their community.

They'd made it out just in time. "My house was already on fire," Thorne says. "So we stood and we watched it burn until I just couldn't watch it anymore."

All across south-central British Columbia, thousands of people found themselves forced from their homes that day as wildfires began raging across the tinder-dry landscape, straight toward towns and ranches and lumber mills. It was a day few would soon forget, but it was only the beginning of British Columbia's summer of fire.

1

AN UNPRECEDENTED YEAR

By the end of 2017, more than 1.2 million hectares of British Columbia went up in flames and more than 65,000 people were forced from their homes. It was the worst fire season in the province's recorded history, and an unprecedented number of people fled their communities.

You would have to travel back in time nearly six decades to find a wildfire season that even comes close to that degree of devastation. The year was 1958, when 855,000 hectares across the province burned, and until now, that was the worst wildfire season on record. To get an idea of just how destructive the blazes were that year, a newsreel from British Movietone News sets the scene with footage of blackened forests and overworked firefighters.

The province was "ravaged by the worst outbreak of forest fires in its history," the narrator informs us. "Magnificent forests of Douglas fir, so long in growing, so quickly destroyed. Millions of acres between the American border and Alaska have gone up in smoke. Forests have been closed to all logging and sawmill operations, throwing many people out of work."

THE WAY IT WAS

In those days, the B.C. Forest Service erected morbid signs at the side of the road after a wildfire had blown through. A giant cigarette was suspended from a gallows by a noose, with a message beneath: "The one who dropped it should also be hanged."

Back then, the province's strategy for fighting wildfire wasn't quite the professional operation it is today. It wasn't uncommon for government officials to pull untrained men from pubs, hand them an axe, and load them onto a school bus headed for the front lines. But the fire season of 1958 was so severe that the forest service was compelled to

(**Opposite**) British Columbia is now more efficient at fighting fires than it was 60 years ago, but the warming climate means it is also more vulnerable to the flames.

experiment with a new technology: repurposed airplanes left over from the war. For the first time ever in the province, four Boeing Stearman biplanes were reconfigured to allow them to drop nearly 700 litres of water or retardant on the flames. That summer, they would log 900 flight hours bombing fires.

A BRAND NEW WORLD

The world has changed a lot since 1958, and so has British Columbia. Fire management has become more efficient and widespread, and "we hadn't seen something over about 400,000 hectares burned since 1961," says Mike Flannigan, a professor of wildland fire at the University of Alberta.

More than three times that area burned in 2017, which makes it a truly unprecedented year. In fact, the area burned was about 10 times the 10-year average for the province.

This exceptional year also welcomed the new largest wildfire on record for the province. Before 2017, the biggest fire ever recorded had been the Kech wildfire of that historic 1958 fire season. It burned 225,920 hectares in the remote northern Kechika Valley that year, more than a quarter of the total area razed in all of 1958, before simply burning out — no firefighters or water bombers necessary.

The Plateau fire of 2017 was an entirely different beast. It grew to well over twice the size of the Kech blaze 59 years earlier — burning a whopping 521,000 hectares of the Chilcotin Plateau, an area nearly the size of Prince Edward Island. The Plateau was created when 19 smaller fires merged together, and it would take more than two months to get it under control. And unlike the Kech fire of 1958, the Plateau wasn't burning through hectare after hectare of remote northern forest. It was threatening several small communities in the Cariboo region of central British Columbia, not more than 60 kilometres from the city of Williams Lake.

(**Right**) B.C. experienced its worst wildfire season ever in 2017, forcing the evacuation of several cities and small communities. The area burned in B.C. was about 10 times the 10-year average for the province.

(**Far Right**) Vast sections of B.C. forests in the Plateau and Cariboo regions were scorched or destroyed by the relentless spree of the 2017 wildfires.

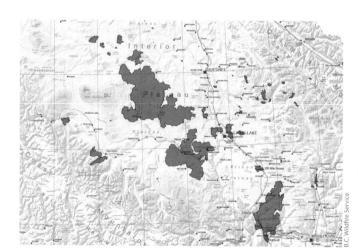

(Far Left) Be hanged: The B.C. Forest Service didn't pull any punches with this sign outside Manning Park, which was erected following the devastating wildfires in 1947.

(Left) Goldstream firefighters fill booster bags and tend pump during 1957 wildfires. In the past, local residents were pulled from their communities to help battle blazes threatening their homes.

MORE FIRES IN MORE PLACES

The fact remains that even if British Columbia is more efficient at fighting fires than it was 60 years ago, it has also become much more vulnerable to the flames. The average annual area burned in Canada has doubled since the 1970s, and that's a direct result of a warming climate.

"Weather is the key driver. You get a warmer world, you're going to have more fire," Flannigan says.

British Columbia's ballooning population only serves to intensify the impact of global warming. About 1.5 million people lived in British Columbia in 1958, but by the summer of 2017, the population had more than tripled to 4.8 million. More people means more speeding trains throwing off sparks, more off-road vehicles with scorching exhaust pipes, more cigarette butts and more neglected campfires. It's a simple formula: more people equals more fire. It also means more homes, more ranches, more oil and gas infrastructure, more cell towers and more remote wilderness lodges, all of which will need to be protected from those fires.

AN UNUSUALLY SNOWY WINTER

Looking back to early 2017, "warming" wasn't a word on many people's minds. The year dawned cold, wet and snowy for most of south and central British Columbia. "If anyone asked me back in February or March if we were going to have any kind of a fire season, I would have said no," says Robert Gray, a fire ecologist based in Chilliwack who has studied wildfires for three decades. January and February were frigid, with lower than normal temperatures across the southern half of the province. By February, the snow and rain had become relentless. Cranbrook, in the far southeastern corner of the province, had nearly four times its usual precipitation that month.

BY THE NUMBERS

The official kickoff of wildfire season in British Columbia every year is April 1st. By the end of that first week in 2017, only one fire had been reported in the province and it had burned just one hectare of land. As the spring wore on, the tally slowly started to rise, and by the first week of May, more than 100 hectares had been scorched in 46 fires. The total hit 824 hectares in the first week of June and then 1,687 hectares by July 1st, but it was still "one of the slower late-spring fire seasons we've experienced," according to Fire Information Officer Ryan Turcot.

(**Right**) The Sikorsky S-55 Helicopter was a staple in the 1957 wildfires.

(**Opposite**) Despite massive fires across B.C., the 2017 fire season was considered "one of the slower late-spring fire seasons" experienced by the B.C. Wildfire Service.

(Opposite) Before the fires came the floods: Firefighter Graham Williams with the B.C. Wildfire Service helped prepare sandbags at one of the many Emergency Sandbagging Stations around Kelowna.

(Left) An extremely snowy winter in Metro Vancouver left road salt in short supply during the spring of 2017. Vancouver residents were so desperate they lined up for hours to get a bucket of salt from Vancouver fire halls.

Even temperate Vancouver experienced a proper Canadian winter, with snow that stayed on the ground for weeks. Road salt disappeared off the shelves of hardware stores, and homeowners lined up for hours outside fire halls to wait for city workers to drop off fresh supplies. By March, as temperatures started to rise, the entire province was drenched. Even traditionally dry places like Kelowna were damp — the Okanagan city experienced its third wettest March on record. In this part of the world, that kind of winter isn't an obvious precursor for a bad wildfire season.

WET WEATHER BRINGS FATAL FLOODS

By early May, all that precipitation meant southern British Columbia was dealing with a different kind of emergency — flooding. Three local states of emergency were declared in the Kelowna area, and about 2,500 people were forced to evacuate across the province. In Cache Creek, Fire Chief Clayton Cassidy was swept away by fast-moving floodwaters; his body wouldn't be found for weeks. Near Salmon Arm, a senior citizen was killed when a mudslide buried his house.

Even wildfire crews, who would normally spend the spring training, setting controlled burns, and clearing brush, ended up on the front lines of the flood response, hauling sandbags and building dams. "When we were in the midst of that flood response, the threat of fire didn't seem all that immediate, but things can change so quickly," says Kevin Skrepnek, British Columbia's chief fire information officer.

(Top) Late Cache Creek Fire Chief Clayton Cassidy was swept away during flash flooding in 2017.

(Top) A wet spring followed by drought conditions in much of the province conspired to provide the perfect combination for a terrible fire season. The Elephant Hill fire, which started near Ashcroft, took months to get under control.

HEATING THINGS UP

Even with all that water everywhere, temperatures were starting to rise, and by June, the rain had essentially disappeared. It's accepted wisdom that a dry June means a bad fire season, and southern British Columbia had a remarkably dry June. In fact, Kamloops saw just 3.4 millimetres of rain in the entire month — less than a tenth of normal levels.

Meanwhile, all that rain and melting snow in the spring had actually helped prime the province for fire. Those damp, fertile fields earlier in the spring allowed for a green explosion at the ground level. Grasses and other low-lying plants proliferated wildly, growing tall and dense and strong. Then, when the rain stopped, those grasses were robbed of moisture, quickly becoming dry and crispy — the perfect fuel for a wildfire. "You just need a good ignition, and if you have some wind events, away you go," says Flannigan.

WILDFIRE SEASON BEGINS

In British Columbia, the shift from one natural emergency to another was sudden. "From one day to the next, we went from flood briefings to wildfire briefings," Environment Canada meteorologist Matt MacDonald says. The first significant wildfire was spotted on June 4th, about 40 kilometres to the northwest of Quesnel. It grew to about 168 hectares, but was easily contained, and no people or properties were ever in danger. Beginning on about June 23rd, new fires started popping up in the Cariboo region, threatening homes and infrastructure alike. An out-of-control blaze near Lac La Hache came within five kilometres of the Esketem'c First Nation's territory, and perilously close to transmission lines. To the southeast, lightning storms were responsible for sparking at least six new fires on June 26th.

By July, parts of the Cariboo were already in extreme drought. Environment Canada weather stations in Kelowna and Victoria didn't measure a single drop of rain the entire month. Because of a quickly growly blaze, the province saw its first wildfire evacuations from campsites near Harrison Hot Springs in the Fraser Valley on July 2nd. The wildfire season was officially underway.

THE EDGE OF A DISASTER

In the days after the Harrison Hot Springs evacuations, residents of more than 100 homes in the Okanagan community of Kaleden fled a fast-moving fire. West of Kamloops, a car crash on the Trans-Canada Highway ignited a blaze that destroyed a woodworking shop and damaged a home. On July 6th, a handful of properties had to be evacuated in the area surrounding 100 Mile House, as the Gustafsen wildfire exploded to 500 hectares in size. By the end of the day, the province had already seen 319 fires that had burned 2,576 hectares of land.

But there was nothing unusual about this. In fact, it was one of the slowest starts to a wildfire season that British Columbia had seen. Unfortunately, luck can all make the difference between just another summer and the worst wildfire season in decades. The perfect conditions for a terrible fire season were already in place — all that was missing was that unlucky element to push things over the edge.

And then came July 7th.

(**Above**) The Cariboo region was one of the hardest hit by the 2017 wildfires, with firefighters using everything they had to battle the hot spots.

(**Above**) As the wildfires swept toward 108 Mile Ranch, Alyssa Procee and Andrea Procee helped wrangle and transport the area's threatened horses to a safe haven in Chilliwack.

2

A DAY TO REMEMBER

It was long past midnight before Angie Thorne got into bed on July 7th. Most everyone else on the reserve was already asleep, but she'd been up for hours, worrying and watching the wildfire approaching from the hills across the highway. She'd seen the trucks from the B.C. Wildfire Service pulling up and the crews spilling out with their gear, and by about 12:30 a.m., the glow in the hills was barely visible.

She figured the firefighters had done their job, and decided to turn in for the night, feeling relatively confident that her home and family were safe. Looking back, she realizes the Elephant Hill wildfire was only biding its time, like some sort of intelligent predator. The fire wasn't done with the reserve. It was just waiting, thinking, planning its attack.

WORST-CASE SCENARIO

July 7th was the day that tens of thousands of lives were put on hold across British Columbia. There would be no lazy July spent splashing in the lake by the beloved family cabin or tending to the vegetable garden or dreaming of a prize pumpkin at the fall fair. Escaping the wildfires with your family intact suddenly became the only thing that mattered.

Before this critical moment, the fires had been relatively manageable but then the weather shifted. After a month of hot, dry weather, bursts of dry lightning struck the desiccated grasslands and heavy winds whipped up the sparks, sending flames speeding across hills and valleys. "If you tried to sit down and plot the worst situation that we could have, that's about what we had," fire ecologist Robert Gray says.

LIGHTNING STRIKES

"Fires are being reported faster than they can be written down," officials with the Cariboo Regional District Emergency Operations Centre warned in a Facebook post that afternoon. Things were quickly spiralling out of control.

(Opposite) Green Lake is a popular hangout in the summer, but the advancing wildfires and thick smoke meant many summer residents had to skip their annual tradition of spending time at the cabin.

(Above) Leslie Huxtable didn't spend much time splashing in the lake or hanging at her cabin on North Green Lake in 2017.

(Right) Langley resident Shawn Cahill was sure his holiday cabin on Loon Lake would be lost as the Elephant Hill wildfire blazed through the lakeside community, leaving little surviving in its wake.

Lightning struck the Cariboo region of central British Columbia 3,305 times in just 24 hours. Across the province, lightning touched down 49,000 times that day — about half the total number of lightning strikes British Columbia typically sees in the entire month of July. One hundred and seven new fires were ignited that Friday and another 68 the next day. No other 24-hour period had more than 40 new fire starts that summer. The fires that were sparked that day weren't just burning through remote forests — many of them were close to communities and getting closer.

It was Hugh Murdoch's 26th season on the job for the wildfire service, and nothing he had experienced in all those years compared to July 7th. He was leading the incident management team for the Gustafsen wildfire, the biggest threat in the province at that point, and was attending a briefing at the Williams Lake airport when all hell broke loose. As he left the meeting and started driving away, he watched four fires break out close by.

"What was so shocking was how these fires took seconds to gain critical momentum. It became immediately apparent that not just our own resources, but the resources of the city in Williams Lake and the community in the surrounding area were going to be taxed — but taxed isn't a strong enough word," Murdoch remembers.

RUNNING OUT OF LUCK

Chief Joe Alphonse had been up early, ready to make the five-hour drive southeast to Kamloops from the Tl'etinqox First Nation's Anaham reserve near Alexis Creek. He was picking up a used hay baler in anticipation of a busy harvest later in the summer. As he drove south down Highway 97 toward 100 Mile House, he could see fires on both sides of the road and started feeling uneasy. "Every summer for the last few years I've been saying we dodged a bullet, and I was thinking, is it this year?" he says.

By the time he reached Kamloops, it was clear the situation was becoming dire. He signed the paperwork as quickly as he could, hooked the baler up to his truck, grabbed a coffee and a sandwich from Tim Hortons, and hit the road. But Alphonse only made it as far as Little Fort, just 93 kilometres north, before he was turned back. In the short time it had taken to finish his business in Kamloops, the fires had already forced the closure of Highway 5.

SCRAMBLING TO SAFETY

Before the day was out, the province had declared a state of emergency that would last for 70 days — more than two months of crisis. "We've been dodging bullets since 2003," says Greg Nyman, who has owned and operated a family ranch near the small town of Clinton since 1987. "Everybody who lives in this area, whose livelihood is the forest and the grasslands, knew that eventually it was going to catch up to us. And boy did it ever catch up to us this year," he says.

Nyman noticed the first plumes of smoke shortly after noon on the 7th. The Elephant Hill fire was still about 30 kilometres south of his ranch, but it was moving fast. He loaded up his kids and a friend and started driving south, to get a better look. They stopped at the Bonaparte Indian Reserve, just north of Cache Creek, where they found most of the community outside, watching the flames to the south.

BY THE NUMBERS

In the week beginning on July 7th, at total of 276 new wildfires were sparked across the province. In that first day of the crisis, the estimated area burned in this wildfire season more than quadrupled, from 2,576 hectares to 11,333 hectares. The estimated total had jumped again by the end of July 8th, hitting 18,241 hectares.

(**Far Left**) Hugh Murdoch, far left, who led the incident management team for the fast-moving Gustafsen fire, said he hadn't experienced anything like it in his 26 years with the B.C. Wildfire Service.

(**Left**) Joe Alphonse, Chief of the Tl'etinqox First Nation, knew it was only a matter of time before a massive wildfire struck the Cariboo.

"Within 10 minutes of us being there, it started burning into the reserve, down off the mountain," Nyman says. "It just jumped from the west of the highway to the east side of the highway. Then everybody was jumping and there was a lot of running and lot of confusion," he says. He helped load up some horses into a stock trailer, and towed the animals away from danger. By the time they were safe, there was no heading back to the reserve — the highways were already closed.

TRYING NOT TO PANIC

To the north, several fires were quickly advancing toward Williams Lake, a city of about 11,000 people and the largest urban centre between Kamloops and Prince George. The fires forced the evacuation of the airport, and about 20,000 people in the city and surrounding area were put on notice that they should be packed and ready to leave.

The city was in panic mode. Evacuees were flooding in from the surrounding communities — signing into reception centres, looking for hotel rooms, and finding room for their livestock at the stampede grounds. Meanwhile, city officials were checking the emergency response plan and preparing for the possibility that everyone would have to leave. "We were just trying to make sure that we had everything in place," Mayor Walt Cobb says. They wanted to be ready to evacuate in case things took a turn for the worse.

FORCED FROM HOME

Even in this record-breaking fire season, nothing even comes close to what happened on July 7th. Firefighters would be chasing the fires that started that day for the next two months. "It was just overwhelming. No fire management agency can deal with that," Flannigan says. As crews struggled to attack blaze after blaze, Mounties were going door to door, ordering people out of their homes. Evacuation orders were issued for the Princeton area, Ashcroft, Cache Creek, 100 Mile House, 105 Mile House, 108 Mile Ranch, 150 Mile House and the Alexis Creek area. By the end of the day, about 3,600 people had been ordered to evacuate communities in central British Columbia. Twelve of the 25 homes on the Ashcroft Indian Band reserve were completely destroyed, and in the Boston Flats trailer park nearby, only one mobile home of about 30 survived the flames.

(**Opposite**) Retardant was dropped near Ashcroft as firefighters tried to get a handle on the Elephant Hill fire.

(**Left**) Greg Nyman refused to leave his ranch south of Clinton despite the heavy smoke and advancing wildfire.

(**Above**) Fire ripped through the Bonaparte Indian Reserve, north of Cache Creek, in early July. Shortly after, Highway 97 was closed and the province declared a state of emergency that would last for 70 days.

(**Above**) Firefighters were deployed across the province, including west of Quesnel, as wildfires continued to spark up.

British Columbia Burning | 23

(Above) All hands were on deck to tackle the Gustafsen fire as it sped along Tatton Road to Lilly Pad Lake Road in the Cariboo.

(Right) Horses belonging to Chris and Dale Watt of Loon Lake found hay and respite at Greg Nyman's ranch near Clinton after their property was evacuated due to the Elephant Hill wildfire.

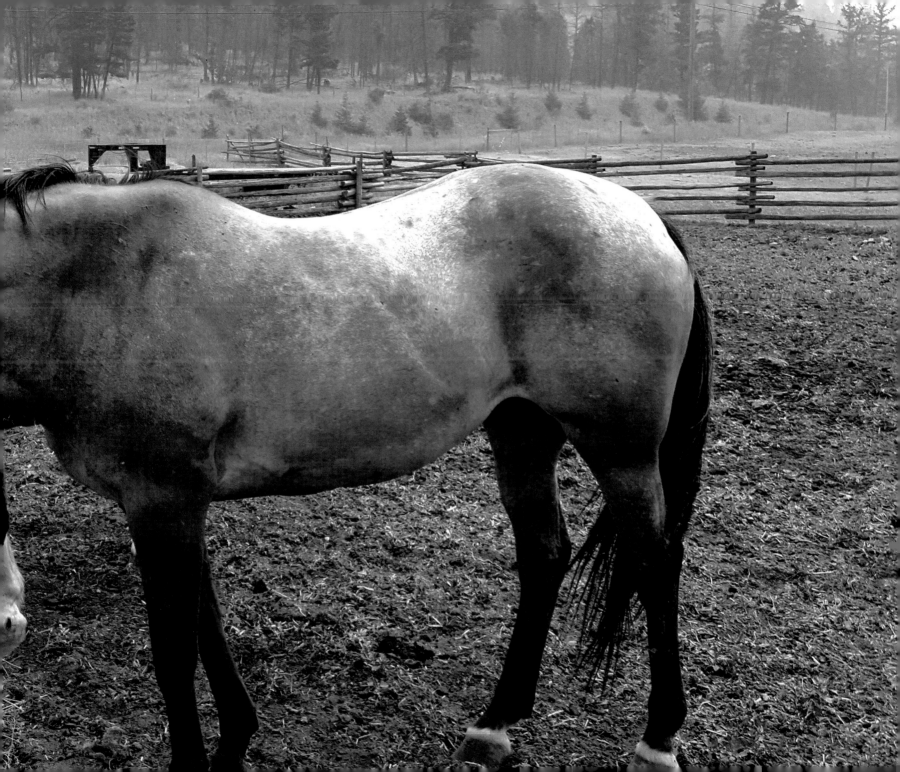

HEART OF THE BATTLE

Once she was certain her house was gone, Angie Thorne helplessly watched the chaos of July 7th play out from a picnic table. "It was like a war zone. There were helicopters flying all over, there were airplanes flying all over, there were bombers, there was a whole mountainside along Ashcroft across from us burning, and we were just sitting there," she says. "There must have been about nine, ten of us, just kind of looking at each other like, asking if that really happened?"

Two days after the wildfire rampaged through the reserve, Thorne was allowed back to take a look at the damage. On the drive in, she passed blackened house after blackened house. "As we rounded the corner and come up upon my place, I think that's when it really hit me — I had no more house there. It was just gone. It was just rubble," she says.

One of her dogs took shelter under a porch and was consumed by the flames. Nineteen of her 21 chickens had perished. "Our two chickens that were left, they ran right up to us, like, get us out of here," Thorne says. "We were just about ready to wrap it up and leave, and then we saw my granddaughter's kitten, running along the side bank....That was a miracle!"

With these small blessings, Thorne and her family returned to their vehicle and drove away from the reserve, heading back to the campground that would be their only home for the next two months. But elsewhere, the trouble was just beginning.

(Opposite) Nevaeh Porter, 8, despairs after seeing the wreckage of what had been her home on the Ashcroft Indian Band Reserve.

(Left) The cat came back: When Kelsey Thorne and her family returned to the Ashcroft Indian Band to survey the damage, they found her daughter's missing kitten running along the bank.

3

THE EMERGENCY ESCALATES

With fires burning to the north around 100 Mile House and to the south near Cache Creek, the village of Clinton was essentially cut off from the outside world from day one of the wildfire crisis. There's only one major road into and out of the tiny community of about 650 people, so it's completely dependent on Highway 97 for the necessities of life.

The source for most of those necessities is Budget Foods, a grocery store and gas station that Jinwoo Kim's family has operated for 20 years. "Nobody was going north, nobody was going south," Kim says. "I couldn't get any trucks in for my supplies. It was just crazy".

"The townspeople needed food, but I couldn't get any food," he says. With his shelves rapidly clearing out and roadblocks on the highway at both ends of town, Kim had only one option: the Pavilion Mountain Road to Lillooet, a 75-kilometre narrow stretch of winding, gravel road. He made the trip twice, taking three pickup trucks with him to meet his suppliers' trucks. "It was crazy. Oh man, I don't want to do that anymore," he says.

LIVING IN LIMBO

By July 15th, eight days into the state of emergency, about 130,000 hectares of British Columbia had already burned — more than had gone up in smoke in all of 2016. In just a week, that number would nearly triple to about 370,000 hectares burned.

The village of Clinton spent these weeks in suspended animation, watching the unpredictable Elephant Hill wildfire approach from the south. Unsure of what was to come, Greg Nyman stayed home from work to tend his ranch outside of town. He was growing increasingly uneasy, believing the provincial wildfire service wasn't doing everything it could to halt the flames. He suspects firefighters missed out on good opportunities to attack the flames on days when the wind was calm.

(Opposite) As wildfires approached from Cache Creek and 100 Mile House, the Village of Clinton was essentially cut off from the outside world.

(**Right**) Budget Foods owner Jinwoo Kim, of Clinton, made harrowing trips along the Pavilion Road to Lillooet to keep the village supplied during the wildfires.

(**Far Right**) Highway 97 through Clinton is usually extremely busy during the summer months, but the village was all but deserted once it closed.

"There just didn't seem to be a lot of concern after it initially burned through Cache Creek and Ashcroft. That's the only place where we saw any bomber activity or any helicopter activity," he says. "I don't think the province was ready for this fire season. They haven't been ready for a fire season for a long, long time."

ONE STEP FORWARD

Over in 100 Mile House, Hugh Murdoch and his firefighting crews were seeing a lot of success fending off the Gustafsen wildfire. When he'd first started working this fire, his boss had impressed upon him just how serious a threat it was to the local community. Two lumber mills were right in the line of fire on the edge of town, and if they burned, it would be a huge blow for people in the surrounding area. In an attempt to protect all those livelihoods, the crews dug machine guards around the mills and set controlled fires on the perimeter, hoping to head the wildfire off at the pass. Sixteen helicopters were on standby and planes were waiting on nearby lakes in case something went wrong.

You can never predict with 100 per cent certainty how a controlled burn is going to behave, but luckily the wind cooperated this time. The plan worked. With the mills protected, there was a real sense of pride, but even then it was impossible to forget that this was just one fire of many, and crews elsewhere weren't having such an easy time of it. Murdoch remembers flying above the fire in a helicopter, and seeing huge columns of smoke coming from the Elephant Hill fire to the south and the dozens of other fires raging throughout the Cariboo. "I think sometimes we get tunnel vision and we think about the situation that's close at hand. We're not privy to the knowledge that there is so much more going on in so many more places that matter every bit as much to the people that live there," he says.

(**Above**) Drew Smith and his roommates loaded up their trucks and headed south when the Gustafsen fire forced the evacuation of their community of 108 Mile Ranch. They stopped at Jake's Pub in 100 Mile House.

ANOTHER STEP BACK

A week into the wildfire emergency, the winds whipped up again, and for a second weekend in a row the province was in crisis. On July 14th, the Elephant Hill wildfire tore through the homes and vacation spots clustered around Loon Lake, 20 kilometres due east of Clinton.

Shawn Cahill, a city firefighter from Langley in the Fraser Valley, stayed behind to try and protect his family's cabin and his neighbours', working tirelessly until the flames drew too close for safety. He remembers the aftermath. "The people all around me lost their places," he says. "You just see the tin roofs laying on the ground where the structure used to be. There was absolutely nothing left, it was so hot."

That weekend, the situation in Williams Lake reached a breaking point. With four large fires closing in and every single evacuation route threatened by flames, the entire city of nearly 11,000 people was forced to evacuate. Thankfully, the municipality had planned ahead for this.

The prospect of a catastrophic wildfire had seemed inevitable since flames came uncomfortably close to the city limits in 2010. By the time the 2017 wildfire season began, staff had detailed plans for an orderly evacuation, and after July 7th, the school board had buses ready at muster stations throughout town. The hospital and seniors' homes had already been cleared out to protect the most vulnerable citizens — the air quality in town had been awful because of all the smoke drifting into town.

BY THE NUMBERS

Though the fires that started on July 7th would continue to occupy firefighters for the entire summer, statistics from the B.C. Wildfire Service show a daily tally of new fires that kept adding to the workload. Sixty-eight new fires started on July 8th, another 27 on July 9th, and then 31 more on July 10th. Throughout that exceptional month, it was rare to see a day with fewer than 10 new fires, and all those blazing days added up to 567 new fires by the end of July.

(Left) When the fire started threatening Shawn Cahill's cabin on Loon Lake, he and his neighbours rallied to save it. John MacIntosh helped by hosing down the area around Cahill's entire cabin to prevent it from catching fire.

THE BEST LAID PLANS

When the order to evacuate came down on July 15th, the plan was to head north to Prince George. Driving south to Kamloops appeared to be out of the question — fires near 100 Mile House had already closed Highway 97 in that direction. "We had everything ready," Mayor Walt Cobb says. "Well, then the fire crossed the road and we couldn't go north." Even though the fires were rushing toward Williams Lake at speeds of up to 40 kilometres per hour when the wind picked up, the city had to put the evacuation on hold as authorities worked out an alternative escape route. Meanwhile, there was always Plan B — gathering everyone into designated evacuation areas at the heart of the city, as far from the surrounding forest as possible, and hoping for the best. "There was no Plan C after that. After that, it would have been helicopters," Cobb jokes.

Thankfully, there was no need for helicopters. The city secured permission to head south on Highway 97, and late in the evening, residents emptied out en masse. It was too late in the day to follow the carefully constructed plan that would have seen neighbourhoods cleared out in stages, so the sudden surge of thousands of vehicles on a single highway created one long, slow traffic jam. "It took 12 or 14 hours to do a three-hour trip because of the traffic," Cobb says.

MASS EXODUS

As all of Williams Lake was heading south, Fire Information Officer Kevin Skrepnek was driving north with a contingent of staff from the provincial wildfire centre in Kamloops, leaving behind a pregnant partner who was nearing her due date. "It was definitely a night I'll never forget," he says. His was one of a handful of vehicles heading into Williams Lake as thousands more streamed out in complete darkness, made even darker by the thick smoke blowing onto the roads from about a half dozen nearby fires.

It was a surreal scene, but the evacuees seemed to be taking it in stride. "Something that really sticks out to me, having driven through that scene, was really how orderly it was — a very Canadian procession," Skrepnek says. "People weren't getting frustrated or angry, we weren't seeing road rage."

Meanwhile, that weekend's dry, windy weather helped spread the trouble to new pockets of British Columbia. Just as Williams Lake was emptying out on July 15th, a nasty grass fire was suddenly ignited in Lake Country, north of Kelowna. Until now, the Okanagan had been relatively untouched by the wildfires to the north and west, but this blaze was ravenous, quickly torching eight homes and damaging more than 30 properties. The fire was almost fully contained within just two days, but the pain it had inflicted was severe, and it was only compounded when RCMP announced that the blaze was the result of arson.

(Above) School buses from Williams Lake were seconded to transport the area's evacuees to Kamloops.

(Bottom) Vehicles left Williams Lake en masse following an evacuation order on July 15.

(Opposite) Residents in the Village of Clinton were constantly on alert as the wildfires threatened them from all sides.

LIFE IN EXILE

By mid-July, Angie Thorne and her family were comfortably settled in the Legacy Park campground in Ashcroft. She was trying to keep a positive spin on things — they had planned several camping trips for their summer vacation, and sure enough, here they were, camping.

Thorne's campsite became the defacto meeting ground for the community displaced from the reserve. Anytime someone needed a break from life in a motel, they'd stop by the trailer for a cold drink, a cigarette and a chat. "I never left my camp for more than one or two days, because what if one of them needs me?" she says.

By July 17th, about 46,000 British Columbians had been forced to leave their homes. With resources strained in Kamloops and the fires along Highway 97 to the north making Prince George inaccessible, communities across the southwest began setting up evacuation centres to welcome the temporarily homeless. Even as far away as the Vancouver suburb of Surrey, a drive of more than 300 kilometres from the main evacuation hub in Kamloops, a hockey arena was equipped with cots and other supplies for evacuees.

But even as those towns and cities filled up with refugees from the wildfires, hundreds of British Columbians were making the tough choice to stay behind despite the danger from fires closing in on their homes.

(Opposite) Wildfire evacuees McLean Rislund, 80, and June Rislund, 81, from Forest Grove near 100 Mile House, walk to an evacuation registration centre in Kamloops.

(Above) Shawn Cahill used to unwind in his backyard forest at Loon Lake before the wildfire struck, scorching his chairs and destroying swaths of trees.

(Left) Floyd Lee, right, cuts watermelon while sitting with his father Garth Lee, after they were evacuated from their homes in 108 Mile Ranch.

British Columbia Burning | 35

4

AN ACT OF DEFIANCE

"You don't say no to an evacuation order just for the sake of saying no," Chief Joe Alphonse says.

For the Tl'etinqox First Nation, the decision to say "no" came after seven hard years of training and planning. The entire community of about 800 or 900 people had been evacuated before in 2009 and then again in 2010. They decided it would never happen again — unless it was on their terms.

It was far too stressful to lie on a cot in an evacuation centre waiting and wondering about their homes as rumours flew. The food in the evacuation centres was unfamiliar to many elders, more accustomed to a traditional diet, and the racism they experienced from some volunteers and fellow evacuees made for painful days and nights.

"We said, we don't want to be a burden on anybody," Alphonse remembers. So when he was stuck in Kamloops on that first weekend of July and heard that RCMP officers were going door to door in his community, telling people to leave, he knew he needed to find a way to get home. It was time to put the plan into action.

BAND WITH A PLAN

Alphonse left Kamloops early on the morning of July 8th and found a route home along deserted highways, arriving in Anaham late in the night. The next day, an RCMP officer came to the band office and told Alphonse to get everyone out. If the children didn't leave, the Ministry of Children and Family Development would come to take them from their parents.

Alphonse was furious. First Nations have the final authority to issue evacuation orders for their reserves, and as chief, he hadn't signed anything. "I told them, 'I understand you have a job to do to inform, you guys have done that, but we are here to tell you guys that we're not leaving,'" he says.

(Opposite) A helicopter provided aid to the Tl'etinqox First Nation, which stayed behind to battle the blaze threatening their reserve.

(Above) People prepared for emergency evacuation from the Tl'etinqox First Nation.

(Above) A firefighter from the Alkali Indian Band gave a thumbs up while walking through a field after finishing a day of putting out hot spots near Alexandria, B.C., where the Soda Creek fire jumped and crossed the Fraser River on July 16.

Every adult who could safely stay behind was assigned a job, and the very young and very old were sent to live with nearby First Nations. Teams of cooks were hired to work around the clock, feeding the crews, and security guards roamed the territory at all hours, looking out for signs of fire and other trouble. Buses were ready and emergency routes were planned in case things took a turn for the worse.

DEFENDING A COMMUNITY

"Every step along the way was a fight," Alphonse says. "We said we're doing this, we're not leaving and we're going to fight these fires, and if you want to be a part of it, then you're going to have to work with us."

They weren't the only Indigenous group to choose this option. Due east of the Anaham reserve, several members of the Xatśūll First Nation also stuck around through the early days of an evacuation order, trying to ward off the wildfire threatening their homes. By the time Williams Lake was evacuated, just 15 kilometres to the south of that reserve, the situation had become too dangerous and the community emptied out.

HOLDOUTS IN A GHOST TOWN

Staying behind after an evacuation order can be an eerie experience. When Mayor Walt Cobb woke up on July 16th, the morning after Williams Lake was evacuated, the city was a ghost town. Just a few hundred people were left — key city officials like Cobb, firefighters, emergency response officials, members of the Armed Forces, and maybe 90 residents who'd refused to leave for one reason or another. Contractors with heavy equipment stuck around to offer their help, and the owner of the local Tim Hortons kept the shop open so the hardworking crews could still have their coffee and donuts.

The city was surrounded by fires, and the streets were filled with smoke. It was hard to breathe and even harder to see. "You couldn't see the hillside around us. To drive downtown and not see a soul was a really, really weird feeling. It just reminded you of one those horror movies," Cobb says. "There were no birds left — the smoke was so bad that the only thing you could hear were crows."

(Above) Members of the Tl'etinqox First Nation preferred to stay and fight the fires rather than fleeing to a shelter, saying if they left they would just worry about their homes.

(Left) The air quality was so bad in Williams Lake that many workers, such as this man at Best Buy Propane, wore masks on the job.

British Columbia Burning | 39

BY THE NUMBERS

By July 29th, just 22 days into the state of emergency, an estimated 427,000 hectares of the province had already burned. In the last decade, British Columbia hasn't come close to that total in an entire season. The largest recent tally was in 2014, when 369,168 hectares were scorched, but the 10-year average is just 154,944.

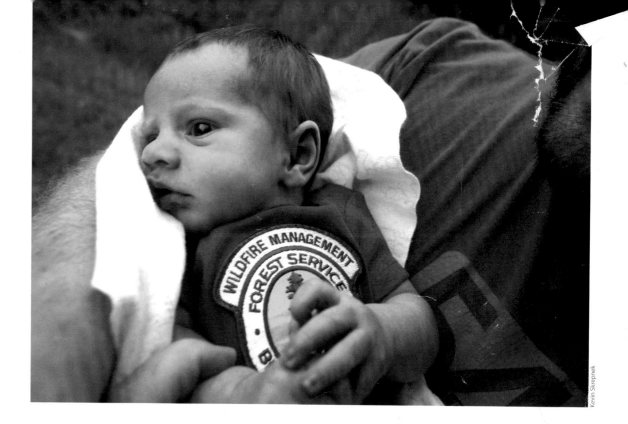

(Right) B.C. Wildfire Service's Information Officer Kevin Skrepnek was able to get back to Kamloops in time to witness the birth of his baby daughter Sage, who was born during the height of the wildfires on July 22, 2017.

(Opposite) Ghost towns and deserted highways, such as Highway 20, became commonplace across B.C. as fires chased people from their communities.

LIFE GOES ON

Information Officer Kevin Skrepnek spent about six days in the empty city, giving reporters daily updates about the fires and the progress crews were making. "It was a bit of a scene, to be honest," he says. "To be in a city, and especially a city of that size, to see it completely empty was quite something. That usual din of civilization was gone." Municipal employees were sleeping under desks in their offices, unable to go home because of the evacuation order.

Skrepnek's heart was back in Kamloops, where his first child was due to arrive at any moment. He made it back just in time for little Sage, a girl, to arrive early in the morning of July 22nd. "I made it back to Kamloops by the skin of my teeth," he says. "I would never have lived it down if I had missed her coming along." Sage made her debut on a Saturday, but by Monday her exhausted dad had to head straight back to work, painfully aware of all the future birthdays he'll have to miss if he continues in the wildfire business.

PROTECTING WHAT MATTERS

For most of those who remained in Williams Lake, staying behind was about carrying out the responsibilities of their jobs. But in communities across the province, ranchers and homeowners defied evacuation orders for reasons of their own, often not trusting the wildfire service to defend their properties and the valuable goods inside. When

arson led to a fire that quickly tore through the Okanagan's Lake Country on July 15th, an art dealer defied the evacuation order so he could protect the irreplaceable paintings he was storing in his home.

The wildfire service works closely with regional governments to determine when it's time to evacuate communities, but Skrepnek has sympathy for the holdouts who stay behind. "I can understand why they do that. I think people's first gut reaction is they want to stay and defend their property, and in a lot of cases it's people's livelihood," he says. "But it obviously puts us in a bit of a tough spot, because we recommend those evacuation alerts and orders for good reason. We don't take it lightly."

BITTER LEGACIES

Being forced to flee can be particularly painful for Indigenous communities, who just happen to be disproportionately affected by wildfire evacuations in Canada. Amy Christianson, a fire social scientist with Natural Resources Canada, studies the impact of evacuations on First Nations people, and she's found that being taken away from familiar cultural practices and territory can be hugely distressing. For many elders, it may be their first time leaving the reserve, and suddenly they're presented with strange foods to eat, new public transportation systems to learn and communities full of people who don't speak their language.

And for many older people, evacuation centres in community halls and hockey arenas bear an uncomfortable resemblance to residential schools. "They're being forced from the community, they're going to stay in communal settings, share bedrooms with people they don't know, have to share washrooms, line up for food — it's the whole thing of not being able to decide for yourself what you want to do."

The residential school legacy was an enormous factor in the Tl'etinqox First Nation's decision to defy the 2017 evacuation order. The experiences of the 2009 and 2010 evacuations had felt unpleasantly familiar — "Again, they're eating in places like a cafeteria, again it's a reminder of residential schools," Joe Alphonse says.

In 2017, there was the threat of getting the Ministry of Children and Family Development involved. Indigenous kids make up close to half the total number of children in foster care in this country, and many Aboriginal parents live in fear of having their children taken away. The Ministry, Alphonse says, "is a real threat in our community and that's past history and that's also current history. It happens more frequently than we would want to admit."

FENDING OFF THE FLAMES

In the end, firefighters from the B.C. Wildfire Service worked alongside about 200 of their Tl'etinqox counterparts. When the nation's firefighters weren't busy protecting the homes of community members, they travelled to nearby ranches to help their neighbours fight off the flames.

(Opposite) A firefighter in Williams Lake took a moment to feed a donkey an apple.

(Above) A log continued to burn on Greg Nyman's rangeland south of Clinton.

(Bottom) When fire broke out in Lake Country, Brian Gervais refused to evacuate his house so that he could protect the art pieces that were stored in his home.

Fires threatened the reserve from the north and then the east, until finally the real threat came from the south, where the flames jumped the Chilcotin River three times. By then, only about 300 community members were left on the reserve — about a third of the usual population. Those people lived through two months of heavy smoke, days when they couldn't see more than 50 metres in any direction even in the light of day. The holdouts spent a week without power, unable to communicate with the rest of the world.

But what really matters is that no one was hurt, and no homes were lost to the flames, although Alphonse jokes that the southern fire had knocked on a few doors, coming within about 100 metres of the community. "If we had followed the evacuation order, we would have lost 30 homes, we probably would have lost our gas bar, our band office, our health office and our church. We would have never recovered," he says.

CUT OFF IN CLINTON

While Alphonse and the Tl'etinqox firefighters were toiling in the smoke, people in the village of Clinton, 150 kilometres to the southeast, would have to wait until July 29th for official word that they needed to evacuate. The Elephant Hill wildfire was still spreading, destroying nearly everything in its path. Store owner Jinwoo Kim was ready to leave with his family, but then his mind turned to all of the friends and neighbours he'd known all his life.

(Above) New Westminster Coun. Chuck Puchmayr stood with Tl'etinqox First Nation Chief Joe Alphonse — in front of the decommissioned fire truck the city donated to the First Nation.

(Above) Greg Nyman was feeling increasingly uneasy as the Elephant Hill wildfire jumped the highway and lit up the hillside near his ranch. Still, he refused to leave.

(Right) Tammy Fisher was surprised to see the smoke clouds as she drove through Clinton ahead of the evacuation order.

(Left) The wildfires took a toll on B.C.'s smaller communities, such as 100 Mile House, as residents fled, leaving few people out and about on the streets.

The owner of the only other gas station in town was already gone, so how would everyone else get enough gas and supplies for the drive out?

Kim lowered all of his prices down to cost, and stuck around to make sure everyone in Clinton was fully supplied. He and his brother Sang sent their employees away. "I was going to leave, but then I saw all the fire trucks — there's about 30 engine trucks come in that night. They came from all over British Columbia, so they're all empty," he says. The Kims spoke with RCMP and firefighters, and everyone agreed the brothers should probably stick around and keep their store open for as long as it was safe.

So Kim packed up his wife and kids and sent them away from town. He kept the store open to supply the crews, while looking out for the few stragglers who had stayed behind, in defiance of the evacuation order. They weren't allowed to leave their homes, so Kim delivered groceries and beer, when he could get it. "I grew up here, so this is my hometown. I didn't want this to be burnt down," Kim says. "This is my life right here. All my friends are here, all my friends' houses are here. For 20 years, they supported me, so I kind of pay them back, whatever I can."

LOSING FAITH

Back on his ranch outside of Clinton, Greg Nyman was defying the evacuation order, hoping to protect his livestock and save as much of his ranch land as he could. As July ended, the Elephant Hill fire was approaching his property from two directions, and he was feeling increasingly nervous. "I'd watched the fire for close to a month and watched the way they'd been fighting it for close to a month, and by then I'd lost all faith in their ability to control it," he says. Things were about to get even worse.

5

WHEN THINGS GO WRONG

As July turned into August, the weather forecast seemed to be stuck on pause. One hot, dry day followed another, with the occasional lightning storm mixed in to make things interesting. An upper ridge of high pressure stayed suspended over most of British Columbia for months — essentially creating a big, warm blob of air. The entire southern half of the province had a hotter and drier summer than normal.

Penticton didn't see a single drop of rain, and in Kelowna, July and August were both hotter and drier than any year on record. The idea of back-to-back months with historical temperature highs and precipitation lows left meteorologist Matt MacDonald feeling a bit stunned. He had to double-check the numbers, saying, "That's incredible. I don't think I've ever seen that."

By August 5th, the province had seen 865 wildfires, burning more than 590,000 hectares. Already it was the second worst wildfire season on record, and it was still the peak of summer.

ANOTHER TURNING POINT

Outside Clinton, the Elephant Hill wildfire continued to frustrate and frighten firefighters. Hot and dry conditions kept it stoked, and the occasional windy day was giving it the strength it needed to continue its assault on the landscape. By now, Hugh Murdoch and his incident management team had moved south to work the merciless fire, stationed at first at a fire camp in Cache Creek. He recalls sitting down for a late dinner at a picnic table one night after a meeting with community members and personnel. The night was dark, smokey and cold — until suddenly it wasn't.

"It got so hot, so fast — hotter than it should ever be at 9:30 at night," he remembers. He looked north, towards Clinton. In his direct line of sight was a field full of tents where the firefighters slept at night. He could see the fabric rippling and gyrating. "It was like a blast furnace had been turned on and pointed right at camp. I was like, oh shit, I gotta go, because I knew I needed to be in Clinton."

(Opposite) By early August, the province had seen more than 865 wildfires, such as this one near Vanderhoof, while more than 590,000 hectares had been burned.

(Above) The tiny community of Lone Butte found itself in the hot seat.

BY THE NUMBERS

By the first day of August, British Columbia had already seen 828 wildfires in a matter of four months, and new ones were still popping up by the dozens every week, and sometimes more than 20 in a single day. In the first 15 days of the month, another 180 new fires had been added to the tally for the year. By August 12th, an estimated 650,000 hectares of the province had burned — more than six times the area scorched in 2016.

A RISKY MANEUVER

Hoping to stop the fire before it burned through Clinton, the B.C. Wildfire Service conducted a series of back burns, setting what were supposed to be controlled fires ahead of the path of the blaze. The idea was to consume all those dry grasses and shrubs before the wildfire could reach them. It's a tricky maneuver that becomes especially risky in mountainous terrain, where canyons and valleys can funnel the wind in unpredictable directions.

"I'd be very, very, very leery or careful to try a back burn in the mountains, unless you were 100 per cent sure of your winds," fire scientist Mike Flannigan says. He was right to be cautious.

By the end of July, it looked to Greg Nyman like the controlled burns outside of Clinton were already failing, allowing the fire to jump across the Bonaparte River and grow even faster than before. He was starting to get concerned about how to save his cattle from the approaching flames. When he heard that a controlled burn was planned near his property for August 1st, he called up local fire officials, asking for a chance to herd his animals away from the range. They gave him five hours.

(Right) B.C. Wildfire Service firefighter Max Arcand uses a torch to ignite dry brush during a controlled burn to help prevent the Finlay Creek wildfire near Peachland from spreading.

(Opposite) The Elephant Hill wildfire got a little too close for comfort for Cache Creek residents and visitors, who were ordered to evacuate.

WORST FEARS REALIZED

"I couldn't sleep all night on July 31st," he says. He was still out on a dirt bike tracking down his cattle when firefighters started lighting up the grassland with their drip torches. The sense of dread intensified. "The wind was blowing in the complete opposite direction for a safe back burn," he says. He watched as a controlled burn became completely uncontrolled, leaping across Highway 97 and up a ridge, gaining at least two kilometres of ground within just half an hour. Soon, it was a raging inferno.

"Where they initiated this burn, three valleys meet and the wind conditions are notoriously volatile. They swirl — you don't know what they're going to do from one moment to the next, especially in the height of summer like this," Nyman says. "They think they knew best, but it was a disaster." Over the next few days, the planned burns continued. "They did back burn after back burn until basically our entire range unit was wiped out." Fifteen thousand acres of land, almost completely blackened. Soon, he and his neighbours were demanding compensation from the province for what they'd lost.

(**Opposite**) Greg Nyman struggled to sleep at night as the Elephant Hill fire continued its rampage through his rangelands, wiping out whole sections of forest and threatening the safety of his cows.

(**Left**) Firefighters face an escalating battle, especially in tinder dry conditions when controlled burns can turn into raging infernos.

(**Right**) Wildfire service officials admit they were unprepared for the devastating effects the wildfire would have on the heavily forested and idyllic community of Pressy Lake.

(**Bottom**) The wildfires burned fast and furious through Pressy Lake, destroying 33 homes and 24 outbuildings.

A CHANGE IN THE WINDS

No one starts a back burn with the aim of letting it get out of control, but by the same token, no one can predict with complete accuracy how a fire will behave. "It's all for the right intentions. People are making decisions on the spot. But sometimes it gets away from you," says Robert Gray, who has been setting controlled burns for decades. The wildfire service says that on August 1st, before the Clinton back burn got out of control, the wind had been favourable and the conditions were right, but then a dramatic shift in the wind grabbed hold of some drifting embers and slingshotted them through the air, across Highway 97.

Fire Information Officer Kevin Skrepnek says that by August, back burns like those used near Clinton were one of the only options firefighters had left to try to get the Elephant Hill wildfire in hand. There were thousands of controlled fires set over the course of the summer as crews struggled to gain some ground. "When you've got a fire of that size, it's not a matter of putting people in front of it or aircraft, or laying hose. It's a natural disaster at that point, regardless of how it started," he says. "Using fire to fight fire is really one of the most effective tools we've got." But he acknowledges that the burns near Clinton didn't go quite as planned, and says the wildfire service needs to do a better job of communicating the necessity of doing this prescribed burning. Perhaps most importantly, no homes in the Clinton area were destroyed or even damaged by wildfire or the back burns, and just one outbuilding was lost.

A SUDDEN RUSH OF FIRE

That wasn't the case in Pressy Lake, 50 kilometres northwest of Clinton, where 33 homes and 24 outbuildings were completely destroyed by the Elephant Hill wildfire. People were evacuated from their homes and cabins along the heavily forested lake on July 29th, but the fire wouldn't reach their homes for another 13 days.

The lingering question is what, exactly, firefighters were doing in the meantime to protect those properties. Pressy Lake homeowners have been demanding answers from the province since they were allowed back in early September to see the devastation, and wildfire service officials have admitted they were caught unprepared for the destructive capabilities of the fire when it hit the idyllic lakeside community.

No sprinklers were set up to soak homes and no other equipment was used to protect people's properties, Kamloops Fire Centre manager Rob Schweitzer acknowledged in a letter to homeowners. On August 11th, the fire was still 15 kilometres away from the community, but a sudden shift in the wind allowed it to make up all that ground and more in less than 24 hours. The flames quickly jumped across the containment lines laid down by firefighters, and by then it was too dangerous to send anyone in to try and protect people's homes. It was a "pretty extraordinary day," says Skrepnek.

In the end, it's impossible to predict how a wildfire is going to behave and how much damage it will cause, or to guarantee that homes and farmland will be protected. In the public conversation, we often talk about fighting wildfire in terms of winning a war, but wildfire is an enemy unlike any human army on Earth, and it requires an entirely different approach.

(Above) A helicopter flies over Clinton's Lagoon Court.

(Left) Homeowners at Pressy Lake questioned whether the province could have done more to protect their properties.

6

HOW TO FIGHT A FIRE

(Opposite) Shawn Cahill borrowed his neighbour's pontoon boat to head out on the water and survey the fire's progress at Loon Lake, fearing his cabin would be among the wreckage.

(Below) Fires are so complex that they are incredibly difficult to contain. When a fire starts candling and the crowns of the trees are fully engulfed, it can spread super fast, while once it finds its way below ground, it can feed on peat and other organic materials for a year.

When you're fighting a wildfire, some mistakes are to be expected. It's a lot more complicated than a kitchen fire — you can't just toss a pot of water on a wildfire and call it a day. Once the flames jump up into the trees or burrow below the ground, a wildfire can be a very difficult beast to tame.

Shawn Cahill got a chance to see the complexity of a wildfire firsthand in 2017. The Elephant Hill wildfire took just a day or two to roar across Loon Lake, destroying 40 homes, the volunteer fire hall, and 32 other structures; but remnants of the fire lingered for much longer.

Cahill, who works full time as a structural firefighter in Langley, spent an entire month driving back and forth to Loon Lake as he tried to keep the flames from doing any more damage. "In the night, we'd go out in the pontoon boat in the dark and look for the glows that were close to the cabins," he remembers.

BURNING ABOVE AND BELOW

The thing about wildfire is that once it finds its way below ground, it can feed on peat and other organic material for months if the weather is dry enough. In some cases, a wildfire can smoulder underground through an entire winter and then reemerge in the spring to wreak more havoc.

The wildfire in Fort McMurray, Alberta, that began in the spring of 2016 wasn't completely extinguished until a full year and three months later, in part because it took refuge below ground when the weather turned cold. In Loon Lake, Cahill saw that even when it looked like the wildfire was out, it was actually still boiling four feet below the surface. "The fire would just find a root and pop up somewhere else. It was pretty intense."

(**Right**) Firefighters on the ground with axes and other hand tools will always be needed to fight wildfires.

(**Far Right**) If a fire gets big and hot enough, it can generate its own weather patterns. At Sheridan Lake, it throws off an eerie glow.

(**Opposite**) The goal in fighting wildfires is containment. Helicopters, skimmers and people on the ground are all needed to help tackle the flames.

BY THE NUMBERS

2017 was the most expensive wildfire season in British Columbia's history, costing the province more than $568 million in firefighting expenses alone. But even in quiet years, battling this natural enemy is costly. The 10-year average for the B.C. Wildfire Service is $182 million — and there hasn't been a year in the past decade when it cost less than $50 million to defend the province from flames.

And then there's the danger from above. Once a fire starts candling and the crowns of the trees are fully engulfed, it can spread so fast and so far that it's difficult to comprehend. "People have a conception of fire as being like a tsunami — just a uniform wall of flame. In a lot of cases, that's not actually where the damage is done," explains Information Officer Skrepnek.

In Fort McMurray, there were cases where sparks or burning embers were carried up to 10 kilometres by the wind, touching down to start spot fires in stretches of forest that had seemed completely safe mere minutes earlier. Even worse, if a wildfire gets big and hot enough, it can actually generate its own weather, creating winds of up to 100 kilometres per hour. The rising heat from a fire combined with these turbulent winds will sometimes create a terrifying tornado-like vortex known as a fire whirl. The Fort McMurray wildfire was large enough to generate its own thunderheads, known as pyrocumulus clouds, and with those came even more lightning, which meant the fire could spread even faster and further than before.

WAGING AN ANCIENT BATTLE

So how, exactly, do you wage a battle against that kind of enemy? Despite the huge technological advances in most aspects of human life, our approach to tackling the wildfire problem hasn't changed much at all over the last century. At its heart, fighting a wildfire is almost a primal thing — women and men with basic tools taking on the elements. "You cannot replace hard work on the ground. The need to have the guys and the gals working with hand tools and hose will never, ever be replaced," longtime firefighter Hugh Murdoch says. Everything else is just kind of peripheral.

The goal in fighting wildfires is containment. If a blaze is out of control, it's usually considered to be completely uncontained. The goal is to get it to 100 per cent containment as quickly as possible, which means that firefighters have gained control over the perimeter of the fire, and any spot fires that might appear on the outside of that perimeter can be dealt with immediately.

MEN AND WOMEN ON THE GROUND

To reach that goal, the B.C. Wildfire Service relies on four types of firefighting teams, each with varying degrees of glamour attached. There are three-person initial attack crews — teams that rush to the scene of a new fire as quickly as possible. Their job is to try to contain the blaze before it gets any bigger than a few hectares. They set up water pumps, remove brush and other fuels from the path of the fire using chainsaws and axe-like hand tools called Pulaskis, and dig fire guards to limit how far the fire can jump.

"Rapattack" crews do much the same thing, but with a lot more flare. They appear by helicopter and rappel into remote forests that are unreachable by road, often clearing pads big enough for the choppers to return and land with equipment and reinforcements. It's a flashy job, but it isn't nearly as dramatic as that of the "parattack" crews that do away with the ropes and jump straight into danger by parachute. They're the famous "smokejumpers" whose dirty, dangerous jobs have been sensationalized in movies and books.

(**Opposite**) The Princeton wildfire contributed to what is officially the worst fire season on record in B.C. As of August 16, close to 900,000 hectares had been burned and the province was still hot and dry.

(**Above**) Fire crews used water pumps and other equipment to try to contain the wildfires before they grew bigger than a few hectares.

(**Left**) Firefighters took advantage of whatever little time they could find for a break from battling the fire at Loon Lake.

British Columbia Burning | 59

(Above) Firefighters took to the skies west of Williams Lake.

(Below) The last drop: Firefighters attack the wildfire near Kleena Kleene.

And finally, there's the real grunt work that begins when a wildfire can't be contained within a few hectares. "If the fire is the size of my office, you can put it out quickly. Once the fire gets the size of a football field, and it's hot, dry and windy, and the fuels are dry conifers, you now have a problem," explains wildfire scientist Mike Flannigan.

At this point, the reins are handed off to 20-person unit crews. They lay lines of hose and pump water from nearby lakes and rivers, chop down trees to create fuel breaks, and use their Pulaskis and shovels to dig hand guards — essentially hand-dug trenches. They'll often have help from bulldozers and other pieces of heavy equipment, which make digging into the ground and knocking down trees to build fireguards a heck of a lot easier.

HELP FROM ABOVE

But a wildfire can quickly get out of control, and when that happens, any sort of attack from the ground is impossible. If the flames are burning at an intensity of more than 2,000 kilowatts per metre — with flames reaching about two metres high — it's no longer safe to have firefighters on the ground. That's when water bombers and air tankers become key. The role of these planes is very specific and somewhat limited.

"Aircraft only buy you time," fire ecologist Robert Gray explains. When a flying tanker drops its payload of water from above, it's simply cooling down the blaze, not putting it out. If retardant is falling, it's meant to slow the spread of the flames, not necessarily to stop them. The chemicals used as retardant and the know-how for dropping it have improved over the decades, but the real job of these impressive flying machines is to make it safe enough down below for the crews to do their work.

Planes and helicopters can also play a key role when fires are first discovered. "Time is so crucial in that first stage of the fire," Skrepnek says. "If we can get aircraft out there early and get retardant around the fire, that can buy us time for our crews to get in on the ground and actually contain it."

LEGENDARY AIRCRAFT

For someone whose community is threatened by wildfire, there is no more reassuring sight than a squadron of planes and helicopters noisily roaring overhead. Even better, a giant water bomber like the Martin Mars.

It's hard to imagine an inanimate object that has gripped the imagination of the British Columbia public in quite the same way as the Martin Mars. Originally built for service with the U.S. Navy in the Second World War, four of these behemoth aircraft were brought to British Columbia in 1959 to be used as water bombers.

They're still the world's largest "flying boats", with the ability to carry a jaw-dropping 27,000 litres of water at a time. The last of these aircraft used to fight British Columbia wildfires was the Hawaii Mars, retired from service in 2013. The emotional attachment to the lumbering red giant persists, and every wildfire season brings another petition to get the plane back to work.

PUNCTURING THE MYTH

The appeal of the Martin Mars is obvious. The more water, the better — right? Ask any seasoned firefighter or wildfire expert about the plane and you'll get a heavy sigh. The very thing that makes the Mars so popular — its gargantuan size — is exactly what makes it an awkward fit for firefighting.

"These big, big aircraft like the Martin Mars that have a big load, you have to pull your resources off the line while they work," Gray says. "It's a heavy load — it'll knock trees over." The enormity of the payload means that ground crews have to pull back so far that whatever time the bomber buys is already mostly spent by the time it's safe enough for firefighters to get back on the line.

But there's more. The Mars can't carry flame retardant, which cuts its potential uses in half. The plane is so big that it can only draw water from about 113 bodies of water in British Columbia, compared to the 1,700 that newer, smaller planes can land on. And in British Columbia's mountainous terrain, its uses are even further limited by the fact that it can only drop its load while flying downslope — it would stall if that happened on an uphill run. "There's just not much use for a big, lumbering, World War II-era museum piece," Gray explains.

Mark Van Manen/Vancouver Sun

TOO HOT TO HANDLE

If a fire is burning hot and wild enough, there is no airplane in the world that's going to do any good. Once the intensity of a wildfire hits 10,000 kilowatts per metre — with flame heights of 50 metres or more and flame temperatures exceeding 1200°C — the crowns of trees are fully engulfed, and even air bombers become totally useless, no matter how reassuring the sight of water bombers might be to those living nearby.

"If it's crowning, it's like spitting on a campfire. It's a nice picture, but it doesn't mean anything. It's not doing anything," Flannigan says.

That's when back burns like the ones in Clinton become absolutely crucial. "In such an extreme year when conditions are so volatile, the land is so dry and the fires are so hot, you cannot get enough water and you cannot get enough people with hand tools in front of any fire," Murdoch says. "We have to fight large fires with fires, and I think that is something that might be catching some of the public by surprise."

(**Above**) The Martin Mars has a huge appeal among the public, but its massive size and inability to carry retardant make it an awkward fit as part of the firefighter tool kit.

7

THE WORST OF TIMES

On August 16th, it was official: after more than 1,000 fires and close to 900,000 hectares burned, 2017 was the worst wildfire season on record in British Columbia. The summer was far from over however; British Columbia was still hot and dry, its fire crews exhausted, its citizens fearful and stressed out.

There were days it really seemed the season would never end. The wildfire service has strict rules about how many days its firefighters can work in a row, but repeated deployments of 14 days straight can leave anyone feeling worn down. "They were working basically as much as they could within the confines of keeping people safe and managing fatigue," Information Officer Skrepnek says. "It was a few days off here and there over the course of the summer, but otherwise it was a full-court press."

BIGGEST WILDFIRE EVER

In August, Hugh Murdoch and his team were sent back to the Cariboo to try and get a handle on the gargantuan Plateau wildfire. By now, 19 smaller fires had joined forces to create this, the largest fire in British Columbia's history. It was so big that it would take a helicopter an hour to fly from one side to the other. At times, it felt like crews were attacking just a tiny pimple on the back of an immense beast of a blaze. At the height of its destruction, 400 firefighters, 75 pieces of heavy equipment and 25 helicopters were all tasked with trying to get it under control.

"We'd run out of radio frequencies. A helicopter goes up in the air and he's catching conversations from six different fires. The volume of work outstretched the infrastructure that we had," Murdoch recalls.

The massive Hanceville-Riske Creek fire burning near Williams Lake ballooned to more than 230,000 hectares in August, making it the second largest in the province. On the Alberta border, the Verdant Creek wildfire in the Rocky

(**Opposite**) Smoke billows near McNeil & Sons Logging Ltd. in 100 Mile House.

(Above) Firefighters set up camp near the Williams Lake Airport.

Mountains west of Banff was an on-again, off-again headache, forcing repeated closures of Highway 93, a key link between the two provinces.

Over in the Kootenays, the McCormick Creek fire was creeping towards the Salmo River Ranch, where 15,000 electronic music fans were gathered for the Shambhala Music Festival on the second weekend of August. The entire four-day event was conducted under an evacuation alert, with just one road connecting the festival grounds to the highway.

Meanwhile, the Elephant Hill wildfire continued its relentless spree across central British Columbia, forcing the province to close off parts of the backcountry to the public. A few days of calm weather might give firefighters the impression they were making progress, only to be followed immediately by bouts of heavy winds, whipping up the flames once again.

FIRE, FIRE EVERYWHERE

With all this destruction in so many places, Murdoch has a hard time picking out one wildfire as the trickiest and most unpredictable. "[Elephant Hill] did some crazy things numerous times. There were fires in the Cariboo, too, that took absolutely incredible, intense runs for extended periods of time. Seven hours of racing through forested areas and just cooking off everything in its path," he says. "It happened in more places than you might think." The good news is that by August 16th the majority of people who were forced to evacuate in mid-July had been allowed to return home, though about 6,000 British Columbians were still living in exile.

In the last week of August, another 1,100 people had to be evacuated east of Kelowna because of the new Philpott Road wildfire. "I remember thinking there are going to be some very, very difficult decisions that are going to have to be made here," Murdoch says. "Resources are going to have to be reallocated, because the province didn't have enough to handle another large campaign fire in the south."

This was one case where firefighters were able to make quick work of the fire, cutting off its growth within about 24 hours and bringing it under control just as the flames reached within a few hundred metres of residential areas

BY THE NUMBERS

When the summer of 2017 officially became the worst wildfire season on record in British Columbia, the burnt area blew the totals for other recent years out of the water. At nearly 900,000 hectares and counting, more of the province had been scorched that summer than the previous five years combined. The worst fire season in the last decade had been in 2014, when 369,168 hectares burned.

(**Left**) The Hanceville-Riske Creek fire burning near Williams Lake ballooned to more than 230,000 hectares in August, making it the second largest fire in the province.

(Right) As residents fled the fires, the criminals slipped into town, staging several break-and-enters and looting homes and businesses. Williams Lake shopkeepers tried to warn off would-be thieves by removing all valuables from their stores.

(Below) Seeing red: Upon returning home to Cache Creek, residents found sections of the town and their possessions covered in retardant.

(Bottom) Police closely monitored who went in and out of B.C. communities, such as Cache Creek, but they couldn't deter the criminal element.

(Opposite) A garden gnome is covered in red retardant at a home in Cache Creek.

A BITTERSWEET HOMECOMING

Residents of Williams Lake were allowed to return to their homes on July 27th, after nearly two weeks in exile. The White Lake fire had come within three kilometres of the city limits to the north, but no homes were touched by the flames. Criminals, though, were another matter entirely. Break-ins were reported in at least three houses while their owners were evacuated, and a couple in their 20s was charged in connection with all of those crimes.

There were other incidents that put the entire city in danger once again. Just a few days after everyone was allowed to return, a drunken 34-year-old man was caught setting off fireworks to celebrate his homecoming. Mounties described it as a careless act that could have led to more wildfires, and he was fined $1,000. "It was frustrating," Mayor Walt Cobb says. "I'm sure it wasn't malicious, it was just not thinking."

The same couldn't be said of five men who were arrested on suspicion of arson in mid-August after several fires were intentionally set in a subdivision on the city's eastern outskirts. Luckily, observant neighbours were on hand to stamp out the flames before they spread or damaged any homes.

EXPLOITING A CRISIS

Though emergency officials have largely applauded the way British Columbians behaved during the wildfire crisis — for staying calm, pulling together, helping friends and strangers alike — the summer also brought out the worst in certain segments of the population. In July, the mayor of Quesnel warned of a phony fire marshal going door to door, warning residents of a nonexistent evacuation alert. The man appeared to be using the fires as an excuse to scope out the contents of people's homes.

In just the first week of the crisis, at least 10 people were arrested after empty homes were looted in 100 Mile House and on the outskirts of Williams Lake. As the summer wore on, scammers were taking advantage of the situation, too. There were reports of fraudsters placing listings for fake rental homes on Craigslist and Kijiji, bilking desperate evacuees out of hundreds of dollars in deposits. There were also cases where firefighters complained that thoughtless summer holiday makers were making their jobs more difficult than they needed to be. Boaters on three lakes in the Cariboo were warned that their watercraft could be confiscated by the RCMP if they didn't get out of the way when air tankers and water bombers swooped in to load up on water.

By the end of August, after nearly two straight months of wildfire chaos, everyone's nerves were frayed. And for those who lost their homes and livelihoods, there was one nagging question: could any of this have been prevented?

8

HOW DID IT COME TO THIS?

Angie Thorne's father, Les Edmonds, has had a lot of time to ponder whether something could have been done long before 2017 to better protect British Columbians and their land from the flames. Since the Elephant Hill fire destroyed his home, Edmonds has been living in a motel room in Cache Creek, where there's not much to do but think. And there's one nagging question that won't go away: If community members on the Ashcroft Indian Band reserve were still allowed to do traditional burning every spring, would they have lost their homes?

For as long as he can remember, Edmonds had gone out every spring to burn sagebrush and grass to make way for wild asparagus. The Nlaka'pamux elder recalls waiting for the rain to end and the winds to blow from the north, away from everyone's homes, before setting the controlled burns.

"There used to be a whole bunch of us would go down and just make a day of it, and after it's all done and the fire is all out and everything, we'd all journey back home and wait for the asparagus to come up," he says. The word he learned for this practice is GwOOypooyemwh (pronounced like woy-BOY-uh-muh-kh).

BURNING TO BRING LIFE

This kind of traditional burning is common among Indigenous groups in Canada. "First Nations communities are highly experienced with fire compared to your average, non-Indigenous community," social scientist Amy Christianson says. Sometimes, fire is used to open up grazing ground for large herbivores, bringing the hunting grounds closer to home. Other times, as with the Nlaka'pamux, it's meant to stimulate growth of food plants like berries or wild asparagus that might otherwise be crowded out.

(Opposite) People returned home to the Ashcroft Indian Band Reserve to find their homes and properties decimated by fire.

(Bottom) In the past, First Nations members would traditionally set controlled burns to stimulate growth of food plants like berries or wild asparagus.

(Right) Prescribed burning used to be much more common in Canada to help clear out the fallen trees, dry brush and other debris that otherwise can become fuel for huge wildfires.

(Opposite) The fire burned so hot and fast on the Ashcroft Indian Band Reserve that not much could be salvaged from the ashes.

But these practices have become increasingly restricted by provincial and federal governments over the last century. Edmonds hasn't burned for wild asparagus on Ashcroft Indian Band lands for at least a decade, after open burning was banned in his area beginning on April 1st of every year. The grasslands just aren't dry enough by the end of March to make for an effective burn, and so the brush is allowed to grow, undisturbed.

The wild asparagus will still make an appearance, but the stalks are spindly, dry and not nearly as delicious. After years without the annual burn-off, Edmonds believes the land around the reserve was ripe for the catastrophic fire it experienced in 2017. "Everything gets a little too bushy, and when there is a fire it's not going in the right direction and it's going toward the houses….That's what happened in this fire this year."

Edmonds lost nearly everything when his home was razed by the Elephant Hill fire. "All we had time for was to run out of the house," he remembers. "I just picked up the cat, ran out to the car and got out of there just in time. We didn't get to save anything, no important papers or any clothing or anything. Just what we had on."

(Right) Firefighters battle a blaze in Vancouver. A not-so-controlled burn was responsible for one of the biggest catastrophes in Vancouver's history. The Great Vancouver Fire of 1886 started with a planned burn to clear brush for the Canadian Pacific Railway.

FIGHTING FIRE WITH FIRE

Edmonds is not alone in thinking that too much of British Columbia was allowed to sit untouched by fire for too long — prescribed burning used to be much more common in Canada. Purposefully setting a blaze under the watchful eye of firefighters and fire behaviour experts helps clear out the fallen trees, dry brush and other debris that otherwise can become fuel for huge, destructive wildfires that nobody can control. Indeed, prescribed fire continues to be a pillar of British Columbia's wildfire prevention strategy, but it's much less common than it once was, and few people have the skill set to perform a safe burn.

In the forestry district near Kamloops, for example, it wasn't uncommon for logging companies to burn 200 or even 300 cut blocks after harvest in a single year. But in the last two years combined, just three blocks were burned. Fire ecologist Robert Gray was supposed to manage four burns across B.C. in the spring of 2017, but none of them happened because the weather shifted so quickly from flooding to fires. More were planned for the fall, but anyone who could have assisted was preoccupied with the wildfires that were still burning.

AN OVERUSED TOOL

To some extent, the current wariness about controlled burns is justified. "There was a lot of prescribed burning done up until about the 1980s — burning done in places where it shouldn't have been. We were actually doing too much prescribed burning," Gray says.

Logging companies looking for a quick cleanup were frequently burning slash to prime their land for tree-planting. It was often unnecessary and it was annoying. In many parts of the country, Canadians started to complain about all the smoke. Wealthy cottagers in Ontario's Muskoka region, for example, lobbied for an end to the burns that were ruining their summer idyll. And there was always the danger that a planned fire might get out of hand.

It's a legitimate concern. The Cerro Grande wildfire of 2000 started with a controlled burn near Los Alamos, New Mexico, and ended up destroying 400 homes and a few structures at the Los Alamos National Laboratory. Scientist Mike Flannigan describes prescribed burning as just one tool in the toolbox for reducing the risk of a wildfire. It needs to be done very strategically, but even with airtight planning, things can go wrong. "You do everything you can, but there's always a risk where something unexpected will occur and you'll have an escape," Flannigan says.
Burning history

A not-so-controlled burn was also responsible for one of the biggest catastrophes in Vancouver's history. The Great Vancouver Fire of 1886 started with a planned burn to clear brush for the Canadian Pacific Railway, just a few months after the city was incorporated. By the time the fire was out, an estimated 1,000 wooden buildings were destroyed. Witnesses said it took just 45 minutes to wipe the city out.

LAGGING PREPARATIONS

Any good toolbox contains several tools, of course. Former Manitoba premier Gary Filmon created an exhaustive list of the wildfire prevention tools available to British Columbia in a report commissioned by the province after the wildfires of 2003 wiped out 334 homes and forced the evacuation of 45,000 people. Filmon had dozens of recommendations for preparing communities for fire, particularly when it came to clearing out potential fuels from the areas surrounding human habitation — something that requires both funding and training.

The provincial government pledged to implement each and every one of those recommendations, and the wildfire service says the Filmon report was instrumental in shaping its current organization, but critics argue implementation has been alarmingly slow. The independent watchdogs at the Forest Practices Board have been tracking British Columbia's progress on meeting the goals set by Filmon for 13 years, and they are not impressed.

"We would like to be able to say that significant progress has been made and the risks are being adequately addressed, but that's not the case," the board's chair, Tim Ryan, wrote in an op-ed in September 2017. Over the last decade, British Columbia has spent an average of $200 million every year to fight forest fires. The budget for preventing those fires almost seems like an afterthought in comparison — about $10 million a year, according to the Forest Practices Board. In 2015, more than a decade after the Filmon report, only 10 per cent of hazardous potential wildfire fuels had been dealt with.

(Top) Fires, such as the Gustafsen wildfire, are never quickly extinguished as they can continue to burn and smoulder underground.

HOW TO PROTECT A COMMUNITY

Much of this has been accomplished through a national program called FireSmart Canada, which brings together homeowners, governments, First Nations, and industry in an attempt to fireproof communities. It covers big-ticket items like fire-resistant building materials for homes and other structures, and it also trains people to keep their lawns nicely mowed and watered, their trees and bushes trimmed, and firewood and propane tanks stored far from their homes. Forests surrounding towns and businesses are thinned so that any fire that arrives will burn less brightly.

A truly FireSmart community might look very different from the wild, forested British Columbia that people who live here hold dear. "I'm a big promoter of golf courses and baseball diamonds," Flannigan says. "Green grass doesn't burn. Sprinklers can be quite effective."

But effectively treating an area to remove fire fuels is hugely expensive — about $10,000 per hectare — and many private landowners simply ignore the FireSmart principles they've learned. Plus, there are limits to what is possible with FireSmart. As Flannigan points out, fires in the coniferous forest can send embers far into the distance, resulting in new spot fires leaps and bounds from the front lines of a blaze "Does that mean you have to remove conifers within two kilometres of every community? That's just enormously expensive," Flannigan says.

INDUSTRY'S ROLE

A great deal of the responsibility for fireproofing British Columbia lies with the forestry industry, traditionally a cornerstone of the province's economy and the backbone of many communities — particularly those in remote, heavily forested areas that face the greatest risk of wildfire. But, according to the Forest Practices Board, the forestry industry has yet to play a significant role in cleaning up dead wood and other fuels in the interface areas surrounding communities.

Too often, slash is left to dry on the ground after logging operations move through; fallen logs make great habitat for salamanders and all sorts of insects, but they also make excellent wildfire fuel.

Meanwhile, logging companies are required to replace any trees they chop down. Tree-planting is now as much of a British Columbia tradition as felling giant firs, and young outdoorsy types often spend their summers hunched over barren hillsides, replacing mature trees with tiny seedlings. But in order to maximize the potential future profits from any given plot, these new trees are usually densely packed together, making them much more vulnerable to fire, according to Gray.

(**Top**) Sheridan Lake is shrouded in an orange glow as B.C. continues to burn.

(**Above**) Pine-beetle stained logs at Quesnel sawmill.

UNINTENDED CONSEQUENCES

It's possible that efforts meant to protect certain endangered species may have made the province even more susceptible to wildfire. The spotted owl, for example, prefers old-growth forests so dense that very little sunlight reaches the forest floor, but Gray believes British Columbia has been too careful about preserving some of those forests. "Between harvesting and wildlife mandates...we've left massive amounts of coarse wood out there," he says. "We have fires that once we get going, they're hard to stop."

(Left) Forests killed by beetles are the perfect fuel for a destructive wildfire.

Then there are the millions of hectares of dead forest ravaged by the mountain pine beetle during an outbreak that lasted from 1999 to 2015. These dried out trees were most volatile in the year after their deaths, when their needles had yet to fall and they could be carried long distances by the wind once they were aflame. Even now, however, the dead trees that are still standing and the ones that have fallen at their feet are the perfect fuel for a destructive wildfire, and create conditions that can be extremely hazardous for firefighters.

NO WILDFIRE PANACEA

If there's a way to make an environment 100 per cent fireproof — beyond drowning it in ocean water or burying it beneath a glacier — we have yet to find it. For instance, Fort McMurray is surrounded by aspen, a tree normally so reluctant to catch fire that it's sometimes called "asbestos forest."

That didn't matter in 2015, when conditions were so hot and dry that fuel moisture levels hit historic lows and wildfire tore through the aspens. In British Columbia in 2017, areas that had burned just two years ago were lit up in flames once again, suggesting that prescribed burning might have done very little to protect some parts of the province.

"There's a point where conditions are so extreme, even if it's a recently burned area, it will race through. Fuel is fuel is fuel," Flannigan says. Those extreme conditions were relentless in 2017, creating a crisis that consumed the province for an entire summer — and even as August ended, the wildfires weren't ready to let up.

9
NOT OVER YET

As summer wound down and Labour Day approached, there was a tentative feeling that things were finally calming down. Most evacuees from the hard hit south-central corridor of the province had been allowed to return home — if their homes were still standing, that is.

But British Columbia was still hot and dry, and when the long weekend hit, the wildfires proved they still had a few surprises left. In the days between August 29th and September 4th, another 76 new fires were sparked across the province. Until then, the very destructive fires had been mostly restricted to a strip of land bounded by the Cariboo in the north all the way down to the U.S. border in the south.

Now, fires were suddenly popping up around communities in the far southeastern corner of the province. Hundreds of homes had to be evacuated in the countryside to the southwest of Cranbrook, along with the popular campsites and beaches in Moyie Lake Provincial Park, as the Lamb Creek fire ran riot across the hillsides. Another wildfire to the northeast of the city prompted the ?aq'am First Nation to empty out its St. Mary's reserve. Directly to the south, a third large fire was burning on both sides of the border with the United States.

"When we got into early September and southeastern British Columbia came into play, that was a challenge, because all of a sudden, an area from Quesnel to Cranbrook — a fourth of the province — was now in play," Fire Information Officer Kevin Skrepnek says. The month of September saw another 190,000 hectares across the province blackened by wildfire, more than the 10-year average for an entire year.

(Opposite) James Thorne (centre) gives his eyes a rub after sitting out in smoky conditions at the Powwow grounds in Kamloops. He left Williams Lake with nearly a dozen of his family members after the city was evacuated.

(Bottom) Popular campsites in Moyie Lake Provincial Park had to be evacuated as fire ripped through the area.

(Right) Mexican firefighters were brought in to aid local firefighters, who were becoming increasingly exhausted as wildfires continued to pop up and burn across the province.

(Far Right) A deer seeks refuge in the burnt forest around Loon Lake.

BY THE NUMBERS

September started out with a bang: a dozen new fires reported on the first day of the month. By now, the province had already seen 1,155 wildfires. By the next day, the wildfire service estimates that more than 1.1 million hectares had already been scorched, a total that hit 1.2 million by the end of the month. Though the weather was cooling down and rain was beginning to fall, another 124 new fires started in September.

DEPLETING FIRE CREWS

Meanwhile, more evacuation orders were being issued in the Cariboo, where the Elephant Hill blaze was still growing. The new school year was supposed to begin on September 5th, but five schools had to delay their openings because they were under evacuation alert.

Meanwhile, many of the young people who are the backbone of the B.C. Wildfire Service had their own classes to think of. Up to 40 per cent of crewmembers, or between 300 and 400 people, had to leave the front lines in September to head back to college. This made support from other provinces and countries, as well as contract firefighters from the forestry industry, absolutely crucial.

For those who remained, fatigue was becoming a real factor. Many of the men and women on the crews had been deployed to one 14-day stint after another since those distant spring days when they were sandbagging during the floods, and the wildfire service was now trying to give them extra time off whenever possible. For Fire Information Officer Skrepnek, life on the road was keeping him away from his newborn daughter, too. "When babies are so new, they're changing so much, I'd be away for two weeks and come back and it was like a whole different creature," he says.

MOTHER NATURE TO THE RESCUE

In the Okanagan, the Finlay Creek fire suddenly exploded over the Labour Day weekend, growing exponentially in just a couple of days from 10 hectares on Saturday morning to 1,000 hectares by daybreak Sunday. Residents of 100 nearby properties in Peachland and Summerland were forced to flee as it closed in.

Hugh Murdoch was supposed to be resting, finally, after a summer of non-stop action, when he got the call asking him if he could handle yet another fire. "You're not going to say no. That's wrong on so many fronts," he says. But unlike the truly savage fires that had struck earlier in the season, this one was contained within two weeks at about 2,200 hectares and without damage to any structures. This time, firefighters had an unexpected ally in another wildfire burning south of the border; it created a ceiling of smoke that trapped moisture in the air from the many lakes of the Okanagan, and lowered temperatures considerably.

By mid- to late-September, fall weather had arrived and the heat was finally off. The provincial state of emergency first declared on July 7th was lifted on September 15th; by then the British Columbia government had been forced to renew it four times. It would take until September 29th for the disastrous Elephant Hill wildfire to be fully contained at nearly 192,000 hectares, though hotspots were still flaring up inside the fire's perimeter. A week earlier, the wildfire service had given its final update on the gargantuan Plateau fire, finally under control at 521,000 hectares.

A TOUGH YEAR FOR WILDLIFE

As exhausted firefighters were mopping up the hotspots from this never-ending wildfire season, the people entrusted with protecting British Columbia's beloved wildlife were also reckoning with the destruction. In the Chilcotin, an entire herd of wild horses was destroyed by the Hanceville-Riske Creek fire. The Orphaned Wildlife Rehabilitation Society in Delta took in three sharp-shinned hawk nestlings that a firefighter rescued from Williams Lake, as well as a young merlin falcon found inside the evacuation zone in 100 Mile House.

(**Above**) Where there's smoke, there's fire, and it seemed that all of B.C. was burning.

(**Left**) B.C. Wildfire Service firefighter Jordain Lamothe took a brief break while conducting a controlled burn to help prevent the Finlay Creek wildfire from spreading near Peachland.

(Right) A baby bobcat was spotted in the ravaged forests around Clinton. The Forest Ministry said it received only a handful of reports of animals dying in the fires but the impacts of the wildfire are likely to be felt for years to come.

(Bottom) A fox has his day in the smoke at Green Lake.

(Opposite) Emma Wedler, 13, was missing home after she and her family fled their property north of Clinton to stay with her grandparents in Lillooet. It took the Wedler family eights hours to drive to Lillooet, with a trailer full of horses and a show steer and a pickup full of kids, dogs, cats and a rabbit.

Chief Joe Alphonse estimates that about 80 per cent of the land under the Tl'etinqox First Nation's protection was burned by the end of the year. Fire crews saw dead moose floating in a nearby lake, apparently killed by exhaustion while they waited out the wildfires. "I've got one picture of a black bear that all the hair's been burnt off. The only hair is on its head. It may have survived the fire, but I don't think it'll survive the winter," Alphonse says.

The Forest Ministry has said it received only a handful of reports of animals dying in the fires, and none of the 40 radio-collared female moose in the Cariboo were killed. But Alphonse maintains that the impacts will be felt for years. In September, he pleaded with the province to call off the moose hunt for the year, saying the loss of tree cover and construction of thousands of kilometres of fire guards had left the animals too vulnerable and their habitat too accessible to hunters.

At first, the government resisted, but by mid-October, the forest ministry had closed parts of the Cariboo to moose hunting. Meanwhile, hunters will now be forbidden from using motorized vehicles in the far southeastern corner of the province from September through June to give wildlife there a chance to recover.

ANIMALS FEEL THE HEAT

The impact of the wildfires on animals could be felt all across the province. The fires ravaged the nesting grounds of many migratory songbirds that make their summer homes in British Columbia, pushing the birds into the surrounding habitat where competition for food and other resources is stiff. That sent many of them flying further west, to the Vancouver area, according to Janelle Stephenson, manager for the Wildlife Rescue Association of British Columbia's wildlife hospital in Burnaby.

"When there's too much smoke and competition for nesting areas, they'll come down to the Lower Mainland," Stephenson says. "They already know the location; they know there's food here; they know there's habitat." Although there's no definitive data linking the rescue group's observations to the wildfires, staff at the hospital noticed a surge in songbirds injured by flying into windows or falling into the clutches of pet cats at the height of the fire season. Between July 1st and October 1st, the hospital took in 1,276 injured songbirds — a 37 per cent increase over the year before.

Stephenson points out that the true impact won't be clear for some time. The fires started during nesting season, which means large numbers of nestlings and fledglings were lost to the flames. Some birds may have put off nesting altogether. "These birds weren't able to reproduce this year, so you'll probably see an impact next year," she says.

(**Right**) The White-winged Crossbill, a rare sight on the West Coast, is cared for at Wildlife Rescue in Burnaby. This redheaded finch's normal migration route sees it through the Interior of B.C. and Alberta, yet this one appeared in the Lower Mainland. This sighting is another clue pointing to the migration changes of songbirds following the devastating wildfires in B.C. in the summer of 2017.

(**Left**) Wildfires provide opportunities for new growth, such as low-lying vegetation like fireweed and mushrooms, and pioneer tree species like red alder and Douglas fir, to make inroads.

WILDFIRES BRING RENEWAL

While some important habitat might have been lost, new habitat will have opened up, too. "It's a normal cycle. Forest fires are actually a very natural thing," Stephenson says. The large trees of a mature boreal forest shut out the sun and hog the soil, making it impossible for smaller plants and shrubs to grow. Wildfire clears the air, so to speak, allowing low-lying vegetation such as fireweed and mushrooms, and pioneer tree species like red alder and Douglas fir, to make inroads. In turn, these plants produce a new variety of seeds and berries that attract new animal life. It's a process known as secondary succession, and it's crucial for biodiversity. "It kills bugs like the pine beetle, and disease. It's great…in terms of biodiversity," wildfire scientist Mike Flannigan says.

Even large mammal species like the Canada lynx, which prefers to make its home in dense, mature forests, can benefit from a turnover in the landscape. Though the woods make for excellent breeding grounds, the best place to hunt for hare and other small animals is in open, shrubby grasslands or young forest. For the lynx, the ideal homeland is actually a mosaic of different ecosystems. Grizzly bears, too, benefit from the fresh crops of berries and exposed roots — and, sometimes, the charred remains of animals killed by a wildfire. The lure of freshly fire-cleared land is so enticing to bison that the Fort Nelson First Nation in northeastern British Columbia uses prescribed burning to lure the dwindling herds back to its territory.

It's true that even in a summer that saw so much destruction, it was possible to see silver linings almost everywhere. And for a few lucky people, the wildfire season of 2017 held small wonders that nearly made up for all the heartache.

10

SILVER LININGS

The last time Ester Spye had seen her cat Socks was July 7th, the day the Elephant Hill wildfire lit up the Ashcroft Indian Band reserve. Spye had to flee her home without much more than the clothes on her back. All the commotion had spooked Socks, and he refused to come out from his hiding place under the bed. A police officer was shouting at Spye to get out, so she gave up on her darling pet, left her doors wide open, and yelled, "Run for it, Socks."

When she returned to the reserve a few days later, her house was gone, and friends found the single blackened bone of a small animal among the ashes. Spye was heartbroken. She imagined her poor cat dying in one small "poof" when the flames tore through the house. Now living in a tiny motel room with the few belongings she had left, Spye spent a good part of each day crying for all she had lost.

A JOYFUL REUNION

Fifty-two days after the fire, Spye's niece asked her to stop by for a visit. It was a bright summer day, and when Spye stepped inside her niece's house, her eyes had trouble adjusting to the dim light. "Look, auntie," her niece said, pointing to a corner. Spye squinted, and there he was: Socks, alive and well. He'd just appeared on the doorstep one day, seeking a familiar face after weeks of wandering, homeless. "He meowed and I picked him up and, ah, what a good reunion. I was so happy to see him," she says.

She brought Socks back to the motel room where she'd been living, and he nervously surveyed his new home, looking into every nook and cranny, sniffing everything in sight. Then, when he was finally comfortable, he started meowing. "He didn't shut up for about a half hour — he was meowing me to death," Spye smiles. "I wish I spoke kitty language. He was really telling me what happened." Though she'd lost almost everything she owned, this small blessing left Spye feeling almost optimistic. "I'd been crying and crying and crying, until I got my kitty cat. That eased up on the tears."

(Opposite) The Elephant Hill fire was moving so fast that Ester Spye didn't have time to grab anything, including her spooked cat, before she was forced to flee her home on the Ashcroft Indian Band Reserve. When she returned, nothing was left but a burnt pole where her shed had once stood.

(Top) Ester Spye had to abandon her cat, Socks, when she fled the Elephant Hill because it wouldn't come out from under the bed where it was hiding. Fifty-two days later, the pair was reunited.

(Right) Shawn Cahill and his neighbour John Macintosh swept up all the dry leaves and twigs, cut the lawn and sprayed the ground around his cabin and the neighbour's to try and prevent the fire from destroying the buildings. The plan worked.

LITTLE MIRACLES EVERYWHERE

Despite the widespread destruction of British Columbia's wildfires, the province escaped serious tragedy in 2017. Not a single human death was recorded over the entire season. Even though so many people lost so much, there were little miracles here and there that made it possible for people like Ester Spye to preserve their optimism.

Much like Spye, Shawn Cahill was absolutely certain about what he had lost. It was his cabin beside Loon Lake, the place where he spent most of every summer with his two teenage children. Back in July, when a neighbour called him at his home in Langley to warn that the Elephant Hill Wildfire was approaching their cabins up at the lake, Cahill dropped everything to drive up to the Interior and try to protect his place. He got rid of all the dry leaves and twigs on the lawn, cut the grass, and sprayed down the ground around his cabin and his next door neighbour's. Then he took out his firefighting gear and got down to business.

As one of a handful of professional firefighters who kept cabins on Loon Lake, Cahill believed it was his responsibility to protect the community. Provincial firefighters were nowhere to be seen, so a handful of homeowners from the Lower Mainland joined members of the local volunteer fire department in trying to fight off the flames.

UNEXPECTED GOOD NEWS

In the early hours of July 14th, Cahill watched what he thought was his cabin go up in flames. The fire was rushing down the hillsides surrounding Loon Lake. Cahill was across the lake from his summer home, trying to fight the flames, when he and his neighbours realized it was too dangerous to stay. They started packing up their things and preparing to flee. "We looked over and went 'Aw, man.' The whole hillside was just orange. That's when we thought for sure the cabin was gone," he remembers. He drove home and broke the bad news to the kids, trying to stay positive — at least no one had been hurt.

A couple of weeks later, he started getting emails from friends who said they'd heard his cabin had been saved. He was sure it was just one of those rumours that circulates after a catastrophe. Then someone sent a photo showing the intact roof of his place peeking through the woods in the background. Still, Cahill remembered that glowing orange hillside and couldn't believe it. It must have been an old shot, he figured.

It was only when Cahill saw the scene firsthand that he realized the impossible had happened — his cabin had been spared. All of the trees around the property had burned, but his place and his next door neighbours' were still there, completely untouched. There was a perfect circle of unburnt land — an exact outline of the area that Cahill had hastily hosed down on the day of the fire. "I was in disbelief. It was unreal," he says. Even now, he has a hard time believing he got so lucky: "Til the day I die, I'll see that cabin blowing orange," he says.

BURNING HISTORY

The legend that's told about the Barkerville fire of 1868 is that it started with a drunken miner who couldn't keep his hands — or his lips — to himself. He tried to kiss a woman inside Adler and Barry's Saloon, but she wasn't having any of it. She managed to fend him off, and he stumbled into a stove pipe, causing the canvas roof to ignite. One hundred and sixteen homes were destroyed in the gold rush town, and it had to be rebuilt from the ground up. A fire brigade was formed to protect the village, but it was mostly destroyed again when a second fire hit the town in 1883.

(**Left**) When it got too dangerous to stay at Loon Lake, Shawn Cahill returned to Langley, fully believing that his cabin had already burned down. But weeks later, he found out it had been spared.

(**Above**) A truckload of dog food was delivered to the evacuation centre in Kamloops.

(**Right**) Volunteers from Kamloops Community Futures fold donated clothes for evacuees at the Powwow grounds in Kamloops.

HUMAN KINDNESS

The good deeds of people like Cahill, who likely saved his neighbour's home as well, can sometimes get lost amidst all the statistics about this unprecedented fire season. "We do start to lose a bit of connection with the human aspect," Fire Information Officer Kevin Skrepnek says. "So many people really sacrificed this year."

Everyone who lived through this fire season has stories about the people who made the days a little brighter. Hugh Murdoch remembers a pair of women in 100 Mile House who took it upon themselves to make grocery runs at least once a day so the heavy equipment operators building fire guards wouldn't go hungry. Residents of Fort McMurray, who know too well the chaos that wildfires can bring, were sending care packages west almost from day one of the crisis.

"It's amazing how people that you don't know can reach out and help out," Chief Joe Alphonse says. He remembers a stranger showing up to the Anaham Reserve with a truckful of meat, just as the Tl'etinqox firefighters were running low on supplies. Volunteer firefighters from other communities dropped by with loads of food and toiletries. A woman who worked for an oil company in Alberta called one day to ask how she could help, and three days later six portable generators appeared.

People across the province were offering up their homes to strangers, volunteering to dish up food or register families at evacuation centres, and mustering their communities to gather donations for evacuees. Mayor Walt Cobb heard story after story about the kindness Williams Lake residents experienced in exile. "There's nothing but good comments about Kamloops, Quesnel and Prince George," he says.

NEW LIFE AMID DESTRUCTION

Evacuee Amy Emery saw just how much difference the right people can make during a crisis. The transplant from Detroit was heavily pregnant with her first child when the Gustafsen wildfire began closing in on the home she and her wife Danielle Biggar had just bought in 150 Mile House. When the crisis broke out on July 7th, Emery was in the Williams Lake hospital with what felt like contractions, trying not to panic as the people around her gossiped about the fire closing in "the 50."

By the time doctors confirmed she was not in labour, Emery and her wife had very little time left to collect their two cats, a dog, and Biggar's two brothers, and begin the drive north. "You could see the fires on the side of the highway and the smoke — it was kind of hard to see where it started and where it ended, if you were driving into something, or if you were not," Emery says.

But almost as soon as they arrived in Prince George, the little kindnesses started pouring in. Someone anonymously picked up the tab for the single motel room the family had crammed into for the first two nights; a stranger reached out through Facebook to give them a free place to stay for weeks; local businesses offered free breakfasts and pedicures.

(Above) Children had lots to choose from as donations of toys and bikes were dropped off at the Powwow grounds in Kamloops.

(Left) As the wildfires raged across the province, babies were being born everywhere. Evacuee Amy Emery thought it would be good to document the birth of her son and other babies born during the wildfires. Some 31 wildfire babies came together for a photo shoot.

(Bottom) Carver Emery was born during the Gustafsen fire. His parents had just bought a home in 150 Mile House when the community was evacuated.

Laureen Carruthers Photography

A COMMUNITY IS BORN

By the time Emery gave birth to a baby boy named Carver on August 10th, her heart was overflowing from the generosity she'd experienced. When she heard that Carver wasn't the only wildfire baby welcomed into the world in an unfamiliar place in a time of crisis, she began reaching out to other new parents among the evacuees.

"I thought we should get these babies together and get a photo. I know I would like a photo for when he gets older, and he can see all these other babies that were born during the same time," Emery says. She planned a shoot with a local photographer.

Thirty-one newborns showed up to get their photos taken, arriving from across the province. "At first, I thought it was going to be six people, and I said when they grow up they can start a rock band. When it turned out to be so many, I was like, they're going to end up being a whole fire department," she says.

The new parents also brought donations for animal hospitals and rescuers who had cared for pets injured or left behind during the wildfires. Since the photo shoot, they've kept in touch, and raised more money to provide a nice Christmas for the family of one of the wildfire babies, that was struggling with bills. "I want to keep this group together," Emery says. "We can take this picture annually."

The wildfires created a community for all these new parents, and good memories they'll never forget. But just like everyone else affected by the fires, they were still dealing with the fallout from the crisis as the year came to a close. The damage tally continued to rise, and it was becoming clear that decimated homes and razed landscapes wouldn't be the only lasting injuries from this wildfire season.

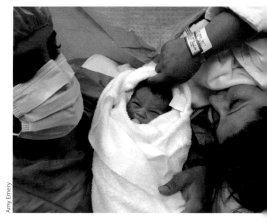

11

TALLYING UP THE DAMAGE

Williams Lake Mayor Walt Cobb gets a bit choked up when he talks about how his city responded to the wildfires of 2017. "I think when it's all said and done, it has brought our community closer together. It has made us realize how important our neighbours are, and how we need to look after each other," he says.

Williams Lake was relatively lucky — no buildings were destroyed or even damaged within the city limits — but rebuilding isn't always a physical thing. The city's fall fair had to be cancelled, and tourism in the region took a massive hit.

Cobb is on the board of Barkerville, a Gold Rush-era historic site to the northeast of Williams Lake, and says revenue from gate fees was down by half a million dollars. The highways were closed for so long that tourists simply couldn't get in. By October, the city itself was still feeling the economic impact of the wildfire season.

"The issue that's around us right now is that there is a number of people that have, for whatever reason, decided not to come home," Cobb says. "Many, many of our small businesses have signs in their windows: Help Wanted. That is something that we never anticipated."

THE HEAVY COSTS OF WILDFIRE

As the season slowly shifted toward winter, British Columbia was picking up the pieces from this unprecedented season. By the end of the year, the estimated cost of fighting the fires of 2017 had hit more than $568 million, and the tally was still rising. About 500 buildings were destroyed, including 230 homes. More than half of those homes were burnt to the ground by the insatiable Elephant Hill fire.

(**Opposite**) Canim Lake was reduced to a barren landscape as a result of the B.C. wildfires.

(**Right**) The wildfires put Williams Lake under extreme stress with evacuations, looting and break-ins. But the community rallied together and gave credit where it was due.

(**Far Right**) It wasn't a rosy homecoming for Cache Creek residents, who found their possessions covered in red retardant.

(**Opposite**) The Boston Flats trailer park was decimated by the Elephant Hill wildfire, with just one trailer surviving the blaze. It was among the carnage of the Elephant Hill wildfire, which the Insurance Board of Canada estimates caused more than $27 million in insured damage.

BURNING HISTORY

Though the damage of 2017 was unprecedented, most British Columbians didn't have to cast their minds back too far to recall another wildfire season with devastating results. The summer of 2003 had been another one to remember, and no fire was as damaging as the Okanagan Mountain Park wildfire, which burned significant portions of both Kelowna and Naramata. A total of 33,050 people had to be evacuated, and about an eighth of those people would be forced out a second time after they were allowed to return home. By the time the fire was out, 238 homes were lost or damaged.

The Insurance Board of Canada estimates that the fires near Williams Lake caused more than $100 million in insured damage. On its own, the Elephant Hill fire dealt a blow of more than $27 million. The Insurance Corporation of British Columbia, the province's public auto insurer, had received more than $860,000 in claims by the end of October. The province's forestry industry will also take a huge hit. The British Columbia government estimates that by mid-December, 26 million cubic metres of green timber had burned — enough wood to build more than 366,000 houses.

A YEAR OF LOSS

The economic impacts extend from the biggest multinational lumber companies to the smallest of small businesses. Jinwoo Kim's grocery store and gas station in Clinton depends on the summer crowds to make ends meet. Adventure-seekers driving up Highway 97 to explore the wilderness and a medieval festival in August help make up for the slow winter months in the isolated village. Almost all that summer business disappeared in 2017, but Kim is just grateful that no one in town lost their home to the fires. "I lost all the summer sales, but you can't be all good, all year. Like every day, it's up and down," he says. "Just no vacation for me this winter. Just gotta control my spending."

Outside of town on Greg Nyman's ranch, up to 30 head of cattle were still missing by the early fall. Though his home and most of his outbuildings were untouched, the fences on the property were reduced to charcoal. His rangeland was scorched, and he estimates it will take years before the grass returns to normal. "There's going to be a lot of people that don't recover from this. It'll be the end of them. They'll either sell out or stop raising cattle," he says. "If you were to ask my wife and my kids, they would say it's time for us to move on as well. Our unit has been destroyed. It won't recover anytime soon."

(Above) Cattle ranchers will likely reassess their livelihoods after wildfire wiped out rangeland across the province, leaving nothing for the livestock to graze.

(Bottom) The 2017 wildfires could be a death knell for ranchers like Greg Nyman, who expects it will take years for his rangeland to recover.

THE WRECKAGE OF FIREFIGHTING

Most of the fences around the Tl'etinqox First Nation's land were lost, too, and the land is now black. The community's firefighters managed to save every home on the reserve, but that meant digging kilometre after kilometre of fireguards — damage that will now have to be repaired. "Now we have way more access into our community than we've ever had," Chief Joe Alphonse says. "If you were to follow the fire guard around the Plateau fire, that's 1,000 kilometres of fire guard…and that's just the Plateau fire." As for his hay harvest? Ruined. He usually makes between 250 and 300 bales in the summer. In 2017, he produced only 66.

There may be lingering impacts on human health, too. By the end of the summer, Environment Canada had issued 33 air quality advisories for the province. Kamloops recorded its worst ever air quality reading on August 3rd — a mind-blowing 49 on the Air Quality Health Index, a system that usually rates the concentration of pollutants on a scale from one to 10. Metro Vancouver hadn't experienced wildfire smoke like this in its recorded history; the regional government issued five advisories over the summer that stayed in effect for 19 days altogether. Smoke from the fires stretched all the way across Western Canada, even creeping into the skies of southern Saskatchewan.

Smokey air can inflame asthma and make breathing difficult for anyone with pre-existing lung conditions, but there is little solid research about the long-term health effects of spending weeks living in a wildfire haze. Still, a 2014 review of two decades of epidemiological studies on the subject showed consistent evidence that exposure to wildfire smoke can increase a person's risk for respiratory and cardiovascular diseases.

THE EMOTIONAL TOLL

Since September, Angie Thorne has been living with her husband and granddaughter in her brother's new house off the Ashcroft Indian Band reserve as they wait for their home to be rebuilt. Her granddaughter, Nevaeh, wasn't able to go to summer hockey camp as she'd planned, but in the fall some cousins dropped off a full set of equipment so she could get back on the ice as soon as the season started. Thorne is happy that Nevaeh is happy, but she also struggles with the idea of accepting help.

"I've always been that person that people can call up, no matter what. Don't get me wrong, I've had the best people that surrounded me and supported me, it's just tougher. I'm the one that's supposed to be standing up on my own," she says.

UNDERSTANDING THE IMPACTS

The psychological impact of evacuations can be particularly difficult. Despite everything that scientists and firefighters have learned about wildfires over the years — how to fight them, how to prevent them — the emotional and social consequences of wildfire are little understood. "This fire season is going to have huge sociological impacts, and we don't have any data on that," fire ecologist Robert Gray says. He's currently working on a project in East Kootenay that attempts to measure the social costs of a wildfire, but funding for this type of project can be hard to come by.

(Left) Payton Thomsas-Thorne and her family were evacuated from Williams Lake. But after only being in Kamloops for a day, she said her eyes were irritated from the thick smoke and she found it challenging to breathe.

Post-traumatic stress disorder is a very real possibility for anyone who's been forced to abandon their home and community to flee an approaching wildfire. Six months after the 2016 Fort McMurray wildfire destroyed more than 2,000 homes, a team from the University of Alberta surveyed 500 residents and found high rates of mental health problems.

Nearly 20 per cent showed signs of generalized anxiety disorder, 15 per cent likely suffered from major depressive disorder, 14 per cent seemed to be abusing alcohol and about 13 per cent appeared to be struggling with PTSD. A longer term Canadian Forest Service study on the impacts of the destructive 2011 Slave Lake wildfire found the community's children were still showing signs of psychological stress a year after the crisis, even if their homes were still intact when they returned.

LINGERING DISTRESS

Just three months after her home was burned to the ground by the Elephant Hill wildfire, Thorne had never felt so helpless. She picked up a cold in September, and it knocked her on her back for two weeks. Her doctor said all the stress had likely left her with a compromised immune system. "I think the last time I was vulnerable was when I was six years old and I was taken away and put in foster care. I've been a fighter all my life, a go-getter all my life, and everything I worked for just went up in flames," she says.

As a social worker, Thorne knows exactly how she would help someone else in this situation, but she just can't figure out how to apply that knowledge to herself. There are publicly funded counselling services available for those affected by the wildfires, but they're usually only available during working hours, when she has her own job to take care of. She hopes to write a book one day about her experiences during the wildfires and get all her feelings down on paper, or maybe create a painting instead. That might be therapy, of a sort.

For now, all she knows is that her life is going to be different. "My husband's like, we need to buy this, we need to buy that, and I said no. We're living simpler lives after this. I find myself crying over stuff that I bought and I worked hard for — that I thought I needed. Now that I don't have it, I know I'm not doing that again," she says

12

FIRES OF THE FUTURE

British Columbia wasn't supposed to burn like this for another three decades. What happened in 2017 looked exactly like what experts had predicted for the year 2050, using computer models based on a conservative two-degree increase in temperature above pre-industrial levels. "Our predictions about 2050 are going to have to change. It's going to look really, really different," fire ecologist Robert Gray says.

Before 2017 had ended, there were grim signs from elsewhere about what the future might hold. In October, just as British Columbia was cooling down, northern California was ravaged by a deadly series of fires that incinerated communities in wine country, killing at least 43 people and destroying 8,900 structures. Just days later, another 44 people died as flames tore through Spain and Portugal. And then in December, southern California was suddenly ablaze with a series of 28 fires, including the largest wildfire in the state's history. The fires forced the evacuation of more than 230,000 people, and more than 1,300 buildings were destroyed.

In comparison, British Columbia got off relatively easy. No one died, and though the fires affected dozens of communities, they singed just a corner of British Columbia. The north, from Prince George on up, was virtually untouched. A vast landscape of boreal forest, filled with highly combustible spruce and lodgepole pine including huge swathes of trees killed by the mountain pine beetle, still sits there, waiting for ignition. And larger urban centres were mostly spared. The Ontario-based Institute for Catastrophic Loss Reduction has been ringing alarm bells since the beginning of the decade about the possibility of places like Victoria, Whistler and Vancouver being touched by disastrous wildfire.

"In many ways, the U.S. wildfire situation is the canary in the coal mine for Canada, an omen portending how and where the wildfire trend will go in this country," the Institute's Glenn McGillivray wrote in The Globe and Mail in November, 2017.

(Opposite) Angie Thorne comforts her distressed granddaughter Nevaeh Porter upon their return to the Ashcroft Indian Band Reserve. Post-traumatic stress disorder is a real possibility for anyone who has had to abandon their homes due to approaching wildfires.

(Right) The psychological impact of evacuations can be difficult for wildfire evacuees. Months after she lost her home, Ester Spye was still struggling to deal with the loss.

(Opposite) The Ashcroft Indian Band will rebuild the homes destroyed in the wildfire but for some, life will never be the same.

THE NEW NORMAL

Gray says British Columbians should be prepared for repeats of the 2017 season within the next decade. The fire season is starting earlier, lasting longer, and blazes are getting bigger and wilder. In a letter to Premier John Horgan, signed by ecologists from the University of British Columbia and the University of Northern British Columbia, Gray wrote: "The extreme wildfire season of 2017 is not an isolated event. It represents the new normal and is part of a global trend of increasing mega-fires with tremendous social, ecological and economic costs."

There's very little controversy about why this is happening among those who've dedicated their lives to studying wildfires. The debate was settled long ago — human-caused climate change is real and its consequences will be catastrophic. Mike Flannigan wrote his very first research paper on wildfires and climate change in 1991, and his most conservative estimates now suggest that Canada will see the expanse of land burned each year double by the end of the century.

More pessimistic models forecast it'll be more like five times the current area burned. "I'm saying climate change has an impact on fire activity already, and it's only going to increase," he says. "Not every year is going to be a bad fire year. There will be regional variations. There will be temporal variations. Some years will be cooler in weather. But overall, we're going to see a lot more fire."

DIRE PREDICTIONS

As Flannigan says, this isn't a prediction for a far-off future. This is happening now. In British Columbia, the wildfire service used to bring in new recruits for training in April and May, but it's now necessary to start them off in February or March. In Alberta, the wildfire season now kicks off on March 1st, a month earlier than before, and in recent years there have been major fires as early as February. In four of the last five years, more than three million hectares of the Canadian landscape was charred by wildfires — that's an area more than half the size of Nova Scotia.

"Nowhere in our record going back to 1918 do we have a five-year period where four years were over three million hectares," Flannigan says. The only year that didn't exceed that standard was 2016, but that was also the year of the Fort McMurray fire — an event that the Insurance Bureau of Canada has named the costliest natural disaster in the country's history.

Recent research suggests that if average temperatures rise by just four degrees before 2080, the average size of wildfires in the southern British Columbian Interior could more than double, wildfire season could become 30 per cent longer, and summer fire intensity could soar by 95 per cent.

MORE LIGHTNING, MORE FIRE

The connection between warmer temperatures and wildfire is pretty intuitive, but climate change also makes the world more susceptible to fires in less apparent ways. Take lightning, which causes a huge number of wildfires every year in Canada. Studies performed in the United States show that for every degree of warming, there's about a 12 per cent increase in lightning strikes. "Everything else being equal, more lightning means more fire. In Canada, about a third to 40 per cent of fires are started by lightning, but they're responsible for about 85 per cent of area burned," Flannigan says. Lightning is even more significant in British Columbia, where it's been responsible for about 61 per cent of wildfires every year, on average.

And just to further inflame the situation, warming temperatures are known to speed up a process known as evapotranspiration — the evaporation of water from land and lakes, combined with loss of water through the leaves of plants. In other words, as temperatures rise, the atmosphere just gets better at sucking the moisture from the landscape. "Our research has found that for every degree of warming, you need about a 15 per cent increase in precipitation to compensate for this drying effect," Flannigan says. "If you look at the models of the future, almost none of the models over any part of Canada have anything close to that increase in precipitation." That means drier grass and trees, which usually results in fires that spread faster and are more intense.

THE FIGHT OF THE FUTURE

As wildfire seasons around the world get longer and fires get bigger and more intense, it will become impossible to fight these blazes using the same tactics we employ today. At the peak of the 2017 fire season, about 4,700 firefighters and other personnel were tasked with trying to get the British Columbia fires under control. They came from every province and territory in Canada apart from Nunavut. They also came from Mexico, Australia, New Zealand and the United States. Even inmates from four British Columbia jails played a role, setting up and dismantling fire camps, repairing hoses, and testing tools for two to eight dollars a day.

BY THE NUMBERS

A large number of fires in any given year doesn't necessarily mean a particularly devastating season — it's the weather and the fuel conditions that really matter. The statistics from 2017 bear that out: there were 1,339 wildfires in British Columbia that summer, with the last two new fires reported on November 7th. That's actually quite a bit lower than the 10-year average of 1,844 wildfires. In fact, it's the third lowest total in the last decade. Compare that to 2008, for example, when just 18,298 hectares of British Columbia burned, but a whopping 1,861 wildfires were sparked.

(**Opposite**) The Elephant Hill wildfire took an emotional toll on Angie Thorne, who vowed that from now on, her family will live a simpler life.

(Right) Wildfires are likely to be the new normal as temperatures continue to rise. To better address the situation, the province has announced an independent review of the provincial response to the 2017 fires, as well as the floods that preceded them.

But experts warn that even this won't be enough people to fight the fires of the future. In 2017, we were already seeing overlaps between wildfire seasons in the northern and southern hemispheres. When those huge brushfires hit southern California in December, Australian firefighters were two weeks into an early and active fire season. If the United States, Europe, Russia, China and Western Canada all have wildfires to battle at the same time, these countries will no longer be able to draw on each other's resources and cooperate as they do now, and assistance from the southern hemisphere will be spread thin.

As Glenn McGillivray wrote in The Globe and Mail: "We can't continue to make suppression the backbone of wildland fire management in this country. We shouldn't, and with a future state that will include more and larger fires on the landscape (and more assets in the way), we won't be able to."

RETHINKING THE BATTLE PLAN

What that means is rethinking British Columbia's wildfire strategy. In December 2017, the provincial government announced an independent review of the provincial response to the 2017 fires, as well as the floods that preceded them. For experts in wildfire science, there is some hope that this catastrophic year could be a turning point for the province — a time when both the government and the people it serves realize that change is necessary. "You have to change public perception. A year like 2017 may be a way at least for British Columbia to say hey, we can do things that will help reduce the risk of having another 2017," Flannigan says.

He wants to see the province focus more on a "monitor and manage" approach to fires as opposed to full-on suppression. That would mean fighting back hard against any wildfires that threaten communities or farms, while taking a hands-off approach to any fires that aren't immediately menacing, running daily analyses based on the weather to determine whether they could become dangerous.

Inside the wildfire service, this is not at all a controversial idea. It's one of five key measures included in British Columbia's 2010 Wildland Fire Management Strategy, and there were several wildfires during the 2017 season that the province chose to observe rather than attack.

Still, some observers are skeptical about how seriously it's been implemented. "You've got to allow fire in the landscape, because you can't prevent it. If you try and stop it all the time, when it does come, it's goes to high intensity extreme fires that you cannot control," Flannigan says.

REIMAGINING THE LANDSCAPE

But strategy isn't the only thing that's going to have to change. British Columbia itself will probably look much different after a few more decades of fire seasons like 2017. The province takes pride in its worldwide reputation as a land of forests, but Gray believes that people who live here will eventually have to say goodbye to a lot of those trees as the forests burn.

(Far Left) With wildfires expected to start sooner and become more intense, it will become impossible to fight these blazes with the same tactics used today. New measures being touted include a "monitor and manage" approach as opposed to full-on suppression.

(Left) At the peak of the 2017 fire season, about 4,700 firefighters and other personnel were tasked with trying to get the British Columbia fires under control, as B.C. bolstered its crews with firefighters from countries like Australia and Mexico. But if those countries are also battling fires at the same time, there may not be people to spare in the future.

The woods are dominated by coniferous trees like pine and spruce that burn very easily and can become involved in extremely dangerous blazes. If we choose to prioritize human safety and economic activity, replanting the forests with those same species might not be the wisest course. "A future with a lot of trees, that's just not in the cards," Gray says.

We can try replacing conifers with deciduous trees like aspen that are a bit more resistant to fire and burn less violently, but if the climate continues on its warming trend, we'll eventually lose those too. "The question becomes, what will replace the forests? Most likely it's something that's very fire prone, shrubs or grass, that can burn even more frequently than trees," Flannigan says. "The type of fuels may change, but the fuels will be there. We have to learn to coexist with fire."

LEARNING TO LIVE WITH FIRE

That's a discouraging thing to consider, but we can also do a much better job of preventing fires by conducting more frequent and widespread controlled burns and clearing out dead wood and debris from the land surrounding homes and industry. Our governments could be more careful about allowing all-terrain vehicles and camping in the wilderness, strategically closing off the backcountry when conditions are ripe for a fire sparked by a tailpipe or cigarette.

We could also improve our ability to tackle dangerous wildfires before they grow out of our control. Flannigan, for one, would like to see the provinces develop more effective early warning systems, using artificial intelligence that

(Right) The Elephant Hill wildfire made a beeline toward the Boston Flats trailer park, scorching a row of fence posts dividing the road outside Ashcroft.

(Far Left) A fire hose played a huge role in saving Shawn Cahill's cabin after neighbours spent hours wetting down the trees and property around it.

(Left) Advance warning systems, using artificial intelligence that sifts through weather forecasts, precipitation records and data on how quickly fuels are drying out, could potentially help B.C. address forest fires before they go out of control.

(Bottom) The future looks grim as wildfires become commonplace in B.C.

sifts through weather forecasts, precipitation records and data on how quickly fuels are drying out. Theoretically, a system like this might be able to predict a day like July 7th on July 4th, allowing British Columbia to move its firefighters to the expected hotspots and to send for reinforcements from Alberta and Ontario before the wildfires even begin.

"There will be some times when you cry wolf and it doesn't materialize, and you spend money and nothing happens. But most of the time, these predictions are made valid by having the right people at the right place," says Flannigan.

BRACING FOR THE WORST

For those towns and cities hit by wildfire in 2017, the question of how to prepare for the future has a special sort of urgency. Williams Lake Mayor Walt Cobb now looks at his city through a new lens. That line of junipers following the driveway of a neighbour's house from the street right to the front door is no longer picturesque — it's more like a trail of breadcrumbs for a hungry wildfire. But in the end, Cobb acknowledges that wildfire is an inevitability for Williams Lake. "You can only do so much. Mother Nature is going to do what it can do. We just have to be prepared."

Within the wildfire service, people like Hugh Murdoch are awaiting the final review of the 2017 season with great anticipation. It's an opportunity to learn from their mistakes and prepare for the years ahead, when fires are only going to get bigger and resources will be spread thinner and thinner.

"We have to make the most of this opportunity to engage people and government at all levels while their interest levels are high, because pretty quickly, people are going to say we dodged that one, that can't happen again," Murdoch says. "But it's going to happen again."

(Above) The Loon Lake community returns home after the Elephant Hill wildfire devastates the lakeside community.